ALTERNATE SOURCES OF ENERGY:

A Bibliography of Solar, Geothermal,
Wind, and Tidal Energy,
and Environmental Architecture

by

Barbara & David Harrah

with a Foreword by
Brent M. Porter

The Scarecrow Press, Inc.

Metuchen, N.J.   1975

ACKNOWLEDGMENTS

We would like to express our gratitude for their assistance to the staffs at the Ferguson Library in Stamford, Connecticut; the Greenwich Public Library; and Library of the State University of New York at Purchase.

We would also like to personally thank Mrs. Judy Koopmann at the S. U. N. Y., Purchase, library and Mr. Eric Wormser for their assistance and advice.

Our thanks, again, to Jim Godfrey, Director of Libraries, Rye Country Day School.

Library of Congress Cataloging in Publication Data

Harrah, Barbara K
    Alternate sources of energy.

    Includes index.
    1. Power resources--Bibliography. 2. Architecture and climate--Bibliography. I. Harrah, David F., 1949- joint author. II. Title.
Z5853.P83H37   [TJ163.2]   016.3337   75-17853
ISBN 0-8108-0839-0

To Our Parents

Joan and Henry Koch

and

Dale and Rosanna Harrah

TABLE OF CONTENTS

FOREWORD

As this bibliographical reference book to alternative sources of energy becomes available, it is important to look ahead for issues which will arise as these energy sources are utilized for power.

Studies performed for the National Science Foundation, private industry and others; research in several fields for over a half century; practical applications by home owners; as well as recent testimony before Congressional committees have established that alternative sources of energy conversion can make a significant contribution to the power supplies of the nation. One of the applications which appears to be ready for widespread adoption at this time is the heating and cooling of buildings using solar energy, as well as the provision of hot water for these structures.

The international implications are of a great magnitude, especially as one examines the building practices and indigenous methods for harnessing natural energy sources of people in many countries. As we examine roof overhangs, wind scoops, natural ventilation, cooking techniques, recreational and social zones, etc., from the viewpoint of many users of the environment, we see that "there is little new under the sun" in their day to day lifestyles. Builders and architects throughout the world know the importance of a good overhang to block summer sunshine and facilitate winter sunshine. The United Nations, in its many publications in the field, and such authors as Bernard Rudolfsky (Architecture Without Architects) and Pearl Buck have shown the international base for energy conservation and utilization. We should anticipate some of the worldwide situations which may result.

The sun is a limitless source of energy, along with geothermal, tidal and the less constant wind energy. These are unlike other sources which are depletable. The four alternative energy sources are interrelated. The sun with Earth's moon establish the system within which the oceans respond.

vii

The sun affects the winds. Its wavelengths within the light
spectrum are constants for man's calculations, manipulations
--and dreams. Its spent radiation, transformed to other
forms of matter and heat, is prevalent throughout our earth.
As we rediscover the chief inexhaustible source of energy in
a post-industrial world, will we run rampant with it? In
utilizing all forms of natural energy we are presented with
trade-offs and long range dichotomies.

Consider the parameters of use

National planning, country by country, will require new
approaches. Conventional land-use planning may be inade-
quate because it is based on exhaustible energy sources and
limited services within a given area. Needed are common-
ality of interest and a less egocentric reflection of an individual citizen and his impact on the destiny of a city or
town. With alternative energy systems, there is a potential
historic reversal: Man may regain his own castle, and do
so at a time when societies are organized as high density
communities and when few men can afford their own castle.
There is the possibility, for instance, of masses of in-
credibly poor residents of Le Corbusier's cities and mega-
structures in India one day possessing a simple solar stove,
water heater, etc.; then, too, there is modern urban man's
use of his city. The latter increasingly regards his sur-
roundings as a fortified environment, a "defensible spaced"
city (see Oscar Newman's findings) and as a place he would
prefer to express manageable or perceptible personal iden-
tity, jurisdiction and direct control over some piece of space
or activity.

As a case in point, consider the urban infrastructure of
Boulder, Colorado. The city has established a "Blue Line."
No municipal utilities can extend beyond this ring. Nature
might be assumed to be temporarily protected. With the ad-
vent of alternative energy systems, and workable subsystems,
DEVELOPMENT may spring up anywhere, certainly beyond
Boulder's own Blue Line. Energy conversion, when integral
to a specific site and in accordance with the determination of
the individual user, will surely alter the theory of a number
of planning professions, the direction taken by governments
--and the individual or collective incentive regarded by the
populace.

Consider new forms of collection

Engineers and scientists conclude that solar energy systems
alone will collect so much energy, that disposing of this
overwhelming supply will become a major challenge. This
disposal is not to be confused with the inherent losses in
conversion and transmission of power. There are standard
losses expected. The initial conversion as well as transmission systems have restraints; there are limited capacities
for handling limitless energy. With the need for excessive
handling of excessive energy, we may find an indirect variant
of POLLUTION FROM OVERABUNDANCE.

This energy overload is not likely with Peter Glazer's concept; however, the manpower and material required to implement his scheme may place impossible demands upon the
earth's resources. He foresees a satellite collection concept
beyond the earth's atmosphere at synchronous altitude (22,000
miles above ground surface) in which an array of photoelectric cells which convert solar energy outside the atmosphere
to electrical energy. This energy is then beamed to earth
by microwave, the beam uninterrupted by the atmosphere.
The national and international implications of this ultimate
form of collection are obvious. Collective expertise and resources of many nations may make the space platform approach
possible with less danger to the biosphere.

Consider energy "intensiveness"

Glazer's concept not withstanding, energy and capital can be
overly concentrated upon one locale or endeavor, a single
facility or a special project, resulting in high economic and
social cost. This overconcentration can result from the often
separate, emphatic actions of those who are most sincere in
their beliefs; bureaucrats, industrialists, small businessmen,
conservationalists, educators, et al. As ECOLOGY became
a household word, and as a period of environmental awareness
peaked in the early seventies, many people called for the re-
pair of the environment at such a high economic and social
cost that much of the improvement effort would establish even
more "energy intensive" processes in our society.

Conversion of energy can be dispersed and yet localized.
Not only individual households, whether within low, medium
or high density areas, but urban complexes too can be sup-
plied with power with less need for transmission of that

power at (1) a personal unit scale, and (2) a metropolitan or
regional scale. A four-story townhouse will need a greater
collection area than present flat plate collectors can provide,
but a series of windmills could be at the summit of the
World Trade Center in New York.

Hopefully we have the options, without further harm to the
present environment and with little depletion of these usable
energy resources. However, there is the potential difficulty
of repetition, when the concentration of energy collection and
its conversion are equally possible. As much energy inten-
siveness as now experienced can result when a society makes
the new technology available to people directly on an individual
basis. The environment--and cities in particular--may either
radically change in form or stabilize its present form by im-
provements from within, expending less total energy. The
trends in renovating old buildings which have sound structures
--and that are so expensive and energy intensive to build
anew--point to the latter. Experiments in desert climates,
on the other hand, suggest bold new structures for mankind.

Finally, consider the pioneers in alternative energy systems
and their entourage

In the United States, the wide disparity between homeowner
oriented systems and city-wide services is paralleled by the
remoteness and location of work in the field of alternative
energy systems. There is ample communication, although a
rather underground, less integrative spirit exists.

At present, pioneers in the field, the academicians and pro-
fessionals in related fields, and industry are largely pursuing
systems for single family, detached dwellings and suburban
or rural facilities. The social response is negligible at this
point, both from the supplier of prototypal systems and the
early users of these systems. Performance criteria on not
only a technological but also a social basis are gradually
being examined. As the need for development of alternative
energy systems becomes apparent to everyone--fuel shortages
and new federal legislation predicating our lifestyles--research
and implementation will likely increase in scale and compre-
hensiveness. The U.S. endeavor thus inherently offers sub-
stantiation of regional aspects of various prototypal systems,
geographic test data, application of science along with indigenous skills, and updating of work by European and Asian
forerunners in the field. The U.S. prototypes also reinforce

international test facilities in that not only small, single-pur-
pose buildings exist but also a broad range of architecture
is imminent; schools, office buildings, mobile and modular
housing units, and residential complexes, for example.

There are three categories of leaders in the field of alternative energy systems. These respective camps address their
work primarily to solar energy utilization.

First, the ranking leaders in the academic community are
George Löf, Director of the Solar Energy Laboratory at
Colorado State University, Ft. Collins, Colorado; Dr. Erich
Farber, Director of the Solar Energy Laboratory, University
of Florida in Gainesville; and Professor John Yellott, College
of Architecture, Arizona State University. For wind and
tidal energy utilization, Professor William Heronemus at the
University of Massachusetts in Amherst is a leading authority
and designer.

Second, the suppliers of components on a production scale in-
clude Pittsburgh Plate Glass, Pittsburgh, Pennsylvania, and
Revere Copper and Brass, Rome, New York. These companies are manufacturing flat plate solar collectors. There
are several small companies producing other equipment and
components for the utilization of alternative sources of energy.

Third, the developers of systems are represented by corpora-
tions: General Electric, Westinghouse, Honeywell, and
Thompson Ramo & Wooldridge. For wind energy, Grumman
Aircraft is the principal developer. For geothermal energy,
the most significant work is being conducted by the Mexican
Government, whose largest single project is located in
northern Mexico where a geologic abnormality produces dry,
relatively noncorrosive steam. For tidal energy, no significant contracts have been awarded in the United States, where-
as the French are proceeding with an experimental tidal
plant.

It is evident that as these individuals and organizations are
joined by those who have recently shown great interest, the
realization of alternative energy utilization--and the impact
upon our lives--has arrived.

BRENT M. PORTER
Co-Head of the Environics Design Studio,
School of Architecture, Pratt Institute, Brooklyn, New York;
and Research Associate, Community Design Associates, N.Y.

INTRODUCTION

The Arab oil embargo caught the major industrialized nations with their diversification down. And as any good ecologist knows, a lack of diversification makes an ecosystem extremely vulnerable: should a blow befall a one-species community and the blow prove harmful to that species, the community may be destroyed because there is no replacement for the lost or damaged element.

With diversification, the ecological community is cushioned with the variegated physical strengths, immunities, mental prowesses, and natural and acquired survival mechanisms of its inhabitants. A blow that is potentially lethal to one may do little or no harm to the others. The community struck falters but rapidly bounces back.

Economics reflects the laws of ecology within the fiscal world. The effect of the embargo is an example. American energy needs are based almost exclusively on oil, and when a significant amount of that oil was removed we became as desperate as an animal that feeds on a one-species diet and wakes one morning to find his meal-ticket species flown the coop. To cover our vulnerability laid bare by the oil crisis, we must emulate in our energy policy the dictates of the physical world. There it is the animal which subsists on various food (energy) sources that survives. When one dish absents itself, another dish remains. By diversifying our economic energy supplies we will not be caught without fall-back sources.

We must avoid, therefore, the trap of viewing any potential energy source as the "answer" to our energy needs-- no matter how promising its present potential. Oil in the 19th century promised a cheap, plentiful, safe source of energy. Its negative polluting aspects were not considered, its exhaustibility shrugged off as a problem for future generations. Then in the 1950's nuclear energy began priming for its eventual ascent once King Oil abdicated. But today the nuclear panacea is posing more questions than it is answering, and so solar energy proponents link their newly rediscovered product with the ultimate answer. Only time, experimentation, and careful evaluation of the results of applied research will prove or disclaim our hopes for solar energy. But for now we should put aside the illusions of single answers for the safety of diversification. Solar power, like wind power, tidal, and geothermal energy, belongs within a well-integrated community of energy sources, adapted for their particular regions, compatible with their immediate environments.

The energy sources discussed in these pages are not new. At the time of the Revolutionary War, Boston powered a mill by trapping the high tide and allowing it to flow out through the mill wheel. The Philadelphia Centennial Exposition on 1876 featured an "amazing new" solar-run steam engine--twenty-three centuries after Archimedes destroyed an entire fleet with sun power. The wind throughout man's and beasts' history has provided a reliable energy source. For men it has powered ships and wind mills; for birds it is an aid in flight. The Romans long before the birth of Christ used geothermal energy to heat their bath water.

Of the four sources of solar, geothermal, wind and tidal energy, solar power holds the greatest promise as a widely usable energy source. More research is being conducted in this area, and more has been written on it. We have, therefore, devoted the greatest part of this bibliography to solar energy and have compiled the significant literature of the last five years as well as the "classics" as a springboard for further intensified solar power applications. Producing reasonably priced solar equipment is the next challenge in solar energy research, and once the federal government outgrows its current infatuation with the bureaucratically entrenched conventional power sources and allocates proper funds for development of cheaper solar energy hardware this challenge will be met. The sun can be a significant, on-line power source by the end of this century--at the latest. For it is no longer a question of whether solar energy is practicable: already throughout the country--not just on the sunny western desert--solar power is providing heat, hot water, and electricity to residential and commercial buildings. The problem remains in making the costs of solar equipment for residential use competitive through mass production so that solar energy becomes an attractive alternative to fossil fuel sources. The equipment to power an American home costs today between $5000 and $8000. With the present high cost of oil, gas, and coal the investment pays for itself within five years. But the initial cash outlay is still beyond the reach of most homeowners.

The bibliography which follows contains more than 1700 entries, mostly on solar energy. But with an eye on diversity, we have included a number of sources dealing with wind, geothermal, and tidal power. The listings are divided into six subject areas:

(1) Various Unconventional Sources of Energy [entries 1-154]. This section includes general articles that cover more than one of the four sources. An example is entry 60, Allen L. Hammond's "Energy Options: Challenge for the Future." It gives a broad overview of the possible sources of power recommended for national future use.

(2) Solar Energy [entries in 1000 and 2000's]. This section includes articles dealing principally with solar energy power. A Solar Energy Subject Index to the various topics within the general subject--collectors, distillation, conversion, etc.--is found at the back of the book. Books listed in this Index under Solar Energy General are either overviews of the broad topic of solar energy or deal with two or more of the narrower topics. References cited under specific topics do not necessarily deal exclusively with that one topic. It is merely the focus of the reference. This index, therefore, should be used as a loose guide.

(3) Geothermal Energy [entries in 3000's] contains articles and books that deal exclusively with geothermal power.

(4) Wind Energy [entries in 4000's]. This section is a short bibliography providing an introductory reference to sources on the generation of electricity through the use of windmills.

(5) Tidal Energy [entries in 5000's] contains 37 entries and is designed to serve as a brief introduction to the use of tidal power.

(6) Environmental Architecture (Climate and Energy) [entries in 6000's]. Application of new sources of energy and conservation of present and potential energy sources has, and will increasingly, influence the design of buildings and of cities. This section contains 22 books and articles which

concentrate on the architectural aspects of climate and general energy usage and gives 48 cross-reference to sources listed under Section 2 (Solar Energy).

The Appendix contains a directory of the periodicals, firms, foundations, organizations, and individuals who are presently working on the application of alternative energy sources. The listing is by no means complete, but it contains the major sources of information.

At the very back of the book is a Key to Abbreviations we have used.

Note: Government reports are followed by their NTIS accession number and are available from the National Technical Information Service, Springfield, VA 22151. These reports are sold at a standard price of $3.00 for reports of 200 or less pages, $6.00 for 201 to 600 pages, and $9.00 for reports exceeding 600 pages. Microfiche copies are available for most documents at a standard price of 95¢ each. Always refer to the accession number provided when ordering material from NTIS.

VARIOUS UNCONVENTIONAL SOURCES OF ENERGY

1 Academy Forum, National Academy of Sciences. Energy: Future Alternatives and Risks. Cambridge: Ballinger, 1974. 202pp.

2 Alfven, Hannes. "Energy and Environment." Bulletin of the Atomic Scientists 18(5): 5ff (3pp.), May 1972.

3 "The Alternatives: R & D for New Energy Sources." Economic Priorities Report 3(2): 23-25, May-June 1972. Graph.

4 "Alternatives to Nuclear Power." New York Times p. 30, Jan. 11, 1971.

5 Anderson, Richard J. "The Promise of Unconventional Energy Sources." Battelle Research Outlook 4(1): 22ff (4pp.), 1972. Photos.

6 "Avoid Air Pollution by Unconventional Energy Technology." Umschau in Wissenschaft und Technik 72(21): 675, 1972 (in German).

7 "Beyond the Energy Crisis." Not Man Apart 1(11): 8-9, Nov. 1971.

8 "Breakdown of Committee Energy Jurisdictions." Congressional Quarterly 32(8): 516-19, Feb. 23, 1974.

9 Breyer, Stephen G. Energy Regulation by the Federal Power Commission. Washington: Brookings Institution, 1974. Refs.

10 Burnet, F. M. "The Implications of Global Homeostasis." Impact of Science on Society 22(4): 305ff (10pp.), Oct.-Dec. 1972.

11 Chalmers, Bruce. Energy. New York: Academic

Press, 1963. 289pp. Photos, graphs, diagrs., index.

12 Clark, Wilson (with David Howell, research). Energy for Survival: The Alternative to Extinction. Garden City, NY: Anchor Press/Doubleday, 1974. 652pp. $12.50. Illus.

13 _____, et al. "How to Kick the Fossil Fuel Habit." Environmental Action 5(18): 3-6, 12-15, Feb. 2, 1974. Drawings.

14 Cohen, Alan S., Fishelson, Gideon, and Gardener, John L. Residential Fuel Policy and the Environment. Cambridge: Ballinger, 1974. $10.50.

15 "Comparing Options for the Technologists; Conservation." Mosaic (National Science Foundation) 5(2) Spring, 1974.

16 "Congress Clears Energy Research Funds." Congressional Quarterly 32(26): 1707, June 29, 1974.

17 Cook, Earl. "Energy for Millennium Three." Aware (20): 3ff (4pp.), Feb. 1973.

18 _____. "Energy in the Transition to the Steady State." Paper presented to AAAS 138th meeting, Dec. 26-31, 1971. Philadelphia, 17pp. Chart, graphs, diagrs.

19 Council on Environmental Quality. Energy and the Environment: Electric Power. Washington: Government Printing Office, 1973.

20 David, Edward E., Jr. "Energy: A Strategy of Diversity." Technology Review 75(7): 26ff (6pp.), June 1973. Graph, table.

21 Davis, David H. Energy Politics. New York: St. Martins, 1974. $3.95.

22 de Onis, Juan. "Energy Crisis: Shortages and New Sources." New York Times p. 1, Apr. 16, 1973; p. 1, Apr. 17, 1973; p. 1, Apr. 18, 1973. Diagr., graph, map, photos.

23 "Development of Non-Nuclear Energy Pushed." Congressional Quarterly 32(27): 1759-61, July 6, 1974. Chart.

24 Driscoll, James G. "A Few Victories in the War for Clean Power." National Observer p. 4, Sept. 6, 1971. Drawing.

25 Dwiggins, Don. The Search for Energy. Chicago: Golden Gate/Childrens, 1974. 80pp. $6.39 PLB. Photos. Gr. 5-10.

26 Energy. Selective bibliography available from the Cleveland Public Library, Cleveland, OH.

27 The Energy Crisis. Bibliography. Send self-addressed stamped (20¢) 9" x 12" envelope for free copy from Publishers' Library Promotion Group, P.O. Box 5925, Grand Central Station, New York, NY 10017.

28 "Energy Crisis: Are We Running Out?" Time 99(24): 49-52, June 12, 1972. Graph, photos.

29 Energy Crisis in America. Washington: Congressional Quarterly, 1975. $4.95.

30 The Energy Directory: A Comprehensive Guide to the Nation's Energy Organizations, Decisionmakers and Information Sources. 1974. Available from EIC Environment Information Center, Inc., Reference Books Div., 124 East 39th St., New York, NY 10016. 500pp. $50.00

31 "Energy from Above and Below." Nature 242(5392): 6-8, Mar. 2, 1973.

32 The Energy Index: A Select Guide to Energy Information Since 1970. 1974. Available from EIC Environmental Information Center, Inc., Reference Books Div., 124 East 39th St., New York, NY 10016. 528pp. $50.00.

33 Energy Policy Project of the Ford Foundation. The Energy Report. Philadelphia: Lippincott, 1974.

34 Energy Primer: Solar, Water, Wind & Bio Fuels. Berkeley, CA: Portola Institute, 1974. $4.50. (pap).

35 The Energy Problem: What Can be Done. Symposium sponsored by the South Plains Section of the Institute of Electrical and Electronics Engineers, the Graduate

School of Texas Tech University, and the Institute for Energy Research of Texas Tech University held February 20, 1974. Lubbock, TX: Division of Engineering Services, Texas Tech University, 1974.   51pp. Illus. , refs. $3. 00.

36 "Energy-Related Research. " Mosaic 5(2): 2-7, Spring 1974.

37 "Energy Self-Sufficiency: What Is the Outlook for 1980?" Congressional Quarterly. 32(23): 1463-66ff 1494-96, June 8, 1974.

38 "Energy Systems: What We Have and Where It Goes. " Mosaic 5(2): 2-7, Spring 1974.

39 "Environmental Exposure to Nonionizing Radiation. " Office of Radiation Programs Report 73-2. Washington: Environmental Protection Agency, May 1973. 143pp. Diagrs. , graphs, maps, tables, refs.

40 Erickson, Edward W. and Waverman, Leonard, eds. The Energy Question: An International Failure of Policy.   Toronto: University of Toronto Press, 1974. 2 vols. $17. 50 each. $6. 50 (pap) each.

41 "Extraterrestrial Imperative. " Bulletin of Atomic Scientists 27(9): 18ff (6pp.), Nov. 1971.

42 Farney, Dennis. "Charles G. Abbot, 99, A Scientist and Sage Looks into the Future. " Wall Street Journal, p. 1, Nov. 3, 1971.

43 Flattau, Edward. "Power: Second in a Series--Is It Necessary?" Science Digest 70(4): 15ff (6pp.), Oct. 1971.

44 Ford Foundation Energy Policy Project. A Time to Choose: America's Energy Future. Cambridge, MA: Ballinger, 1974. $10. 95.

45. "Forum: Energy's Impact on Architecture. " Architectural Forum 139(1): 59ff (16pp.), July-Aug. 1973.

46. Freeman, S. David. The New Era. New York: Vintage/Random House), 1974. $2. 45.

47 Friedlander, Gordon D. "Energy: Crisis and Challenge."
    IEEE Spectrum 10(5): 18-27, May 1973. Diagr.,
    graphs, photo., table.

48 Frost & Sullivan, Inc. Government Energy R & D Mar-
    ket. New York: Frost & Sullivan, Inc., 1974. 309pp.
    Illus.

49 Gannon, Robert. "Atomic Power: What Are Our Alter-
    natives?" Science Digest 70(6): 18ff (5pp.), Dec.
    1971. Photos.

50 Garvey, Gerald. Energy, Ecology, Economy: A Frame-
    work for Environmental Policy. New York: Norton,
    1972. 235pp. $8.95, $2.75 (pap). Gr. 11 up.

51 Georgescu-Roegen, Nicholas. "Economics and Entropy."
    Ecologist 3(3): 13ff (6pp.), July 1972. Drawing.

52 Goldberg, Michael A. "The Economics of Limiting
    Energy Use." Alternatives 1(4): 3ff (11pp.), Summer
    1972.

53 Gray, John E. Energy Policy--Industry Perspectives.
    New York: Ballinger, 1975. $8.50. $2.95 (pap).

54 Grayson, Leslie, E., ed. Economics of Energy: Read-
    ings on Environment, Resources, and Markets.
    Princeton, N.J.: Darwin Press, 1975. Refs. $7.50
    (pap).

55 Grey, Jerry. "Prospecting for Energy." Astronautics
    and Aeronautics 11(8): 24ff (5pp.), Aug. 1973. Draw-
    ings, Refs.

56 _____. The Race for Electric Power. Philadelphia:
    Westminster, 1972. 125pp. Photos., diagrs., index.

57 Halacy, D. S., Jr. The Energy Trap. New York: Four
    Winds/Scholastic, 1974. 192pp. Photos. Gr. 7-12.
    $6.95.

58 Hamilton, [U. S. Rep.] Lee H. "The Energy Crisis: The
    Persian Gulf." Vital Speeches 39(8): 226-31, Feb. 1,
    1973.

59 Hammond, Allen L. "Energy and the Future." Science
    179 (4069): 164-66, Jan. 12, 1973.

60 _____. "Energy Options: Challenge for the Future."
Science 177(4052): 875-77, Sept. 8, 1972. Charts.

61 _____, Metz, William D., and Maugh, Thomas H.,
II. Energy and the Future. Washington: American
Association for the Advancement of Science, 1973.
184pp. Illus., refs. $7.95.

62 Hauser, L. G. "Assessing Advanced Methods of Generation." Electrical Review 193(7): 219ff (4pp.), Aug.
17, 1973. Diagrs.

63 Hellman, Hal. Energy in the World of the Future. New
York: J. Evans & Co. / Lippincott, 1974. Illus.
(Juv.) $5.95.

64 Henderson, Michael D. and Johnson, R. Curtis. "Energy
Conversions and Environmental Pollution." Science
Teacher 39(3): 31ff (5pp.), March 1972. Graphs.

65 "Here Is a Brief Rundown on Ways Electricity is Generated." Aware (27): 3ff (7pp.), Dec. 1972.

66 Heronemus, William E. "Alternatives to Nuclear Energy." Catalyst for Environmental Quality 2(3): 21-
25, 1972. Drawing, photo.

67 Holdren, John. "Defusing Old Smokey by Plugging Into
Nature." Sierra Club Bulletin 56(8): 24ff (4pp.),
Sept. 1971. Drawing.

68 Hollomon, J. Herbert and Grenon, Michel. U.S. Energy
Research and Development Policy. Cambridge, MA:
Ballinger, 1975. 192pp. $11.00

69 Hooker, C. A. "Pollution: The Tip of the Iceberg."
Alternatives 1(1): 35, Summer 1971. Photo.

70 "House Approves Non-Nuclear Energy Research." Congressional Quarterly 32(38): 2547-48, Sept. 21, 1974.

71 Hubbert, M. King. "Energy Resources for Power Production." Paper presented at IAEA/USAEC Conference on Environmental Aspects of Nuclear Power Stations, New York City, Aug. 10-14, 1970.

72 _____. "The Energy Resources of the Earth." Sci-

entific American 224(3): 60ff (11pp.), Sept. 1971. Charts, diagrs., photo.

73 _____. "Survey of World Energy Resources. " Canadian Mining and Metallurgy Bulletin 66(735): 37ff (17pp.), July 1973. Diagrs., graphs, tables.

74 "Improvising Sophisticated Energy Sources May Answer Power Problems. " Commerce Today 3(9): 4ff (5pp.), Feb. 5, 1973. Drawings, photos.

75 Intersociety Energy Conversion Engineering Conference, 9th San Francisco, 1974. Ninth Intersociety Energy Conversion Engineering Conference proceedings, San Francisco, California, August 26-30, 1974. New York: American Society of Mechanical Engineers, 1974. 1343 pp. Illus., refs., indexes. $70.00.

76 Israel, Elaine. The Great Energy Search. New York: Julian Messner and Simon & Schuster Juveniles, 1974. 64pp. Photos., index. Gr. 4-6. $5.29 PLB.

77 Jackson, Frederick R. Energy from Solid Waste. Park Ridge, N.J.: Noyes Data Corp., 1974. 163pp. Illus. $24.00. (Energy Technology Review, No. 1 and Pollution Technology Review, No. 8).

78 Kent, M. F. "Power Generation Options for the Eighties and Nineties. " Public Utilities Fortnightly 90(4): 17ff (5-1/2pp.), Aug. 17, 1972. Drawings.

79 Klaff, Jerome Z. "Power Policies. " Public Utilities Fortnightly 92(6): 31-34, Sept. 13, 1973. Drawing.

80 Kluepfel, Neil. "Out of the Shadow. " Ecology Today 2(1): 29ff (2-1/3pp.), March-April 1972. Drawings.

81 Knowles, Ruth Sheldon. The Energy Famine: What Americans Must Do to Become Self-Sufficient. New York: Coward, 1975. 256pp. Illus. $7.95.

82 Kupperman, Robert H. "Energy Supply Not Keeping Pace with Demand. " Mining Congress Journal 59(3): 25ff (4pp.), March 1973.

83 Landsberg, Hans H. "Low-Cost Abundant Energy: Paradise Lost?" Science 184(4134): 247-53, April 19,

1974. Charts, photo., drawing, refs.

84 _____, et al. Energy and the Social Sciences. Washington: Resources for the Future, 1974, viii, 778pp. (RFF Working paper; EN-3). Refs., index. $7.50.

85 "Lighting Some Candles." Architectural Forum 139(1): 89ff (10pp.), July-Aug. 1973. Drawing, photos.

86 Limaye, Dilip R., ed. Energy Policy Evaluation. Lexington, MA: Lexington Books, 1974. $14.00.

87 Lindsley, E. F. "New KROV Engine Boosts Hope for Steam Cars." Popular Science 202(6): 90-93, June 1973. Diagrs., photos.

88 McCallum, John and Faust, Charles L. "New Frontiers in Energy Storage." Battelle Research Outlook 4(1): 26-30, 1972. Chart, photo.

89 McCaull, Julian. "Oil Drum Technology." Environment 15(7): 13-16, Sept. 1973. Photos.

90 McCormack, [U.S. Rep.] Mike. "We Must Explore New Methods of Energy Conversion." Public Power 31(1): 20-24, Jan.-Feb. 1973. Diagr.

91 McCracken, Paul W. (moderator). The Energy Crisis. Panel discussion by Clifford P. Hansen and others. Washington: American Enterprise Institute for Public Policy Research, 1974. 110pp. $2.00(pap).

92 McLean, John G. "New Year's Resolutions for the Energy Crisis." Drilling Contractor, Jan.-Feb. 1972.

93 Makhijani, A. B. and Lichtenberg, A. J. "Energy and Well-Being." Environment 14(5): 10ff (9pp.), June 1972. Charts, drawing, photos.

94 Mancke, Richard. The Failure of U.S. Energy Policy. New York: Columbia University Press, 1974. $10.00. $2.95 (pap).

95 Massachusetts Institute of Technology. Energy Laboratory. Policy Study Group. Energy Self-Sufficiency: An Economic Evaluation. Washington: American Enterprise Institute for Public Policy Research, 1974. 89pp. $3.00 (pap).

96   "Meeting the Energy Dilemma." New York Times. p.
      32 (editorial), July 2, 1971.

97   Merrill, Richard, et al. Renewable Sources: An Energy
      Primer. Berkeley, CA: Portola Inst., 1974. $4.00.

98   Millard, Reed. How Will We Meet the Energy Crisis?:
      Power for Tomorrow's World. New York: Julian
      Messner, 1971. 189pp. Photos, diagrs., index.

99   Nakamura, Leonard I., ed. Energy Forum (proceed-
      ings), New York: The Conference Board, 1974.
      31pp. $5.00.

100   Nash, Barbara. "Energy from Earth and Beyond."
      Chemistry 46(9): 6ff (4pp.), Oct. 1973. Diagrs.,
      map, photo.

101   National Academy of Engineering, Task Force on Energy.
      U.S. Energy Prospects: An Engineering Viewpoint.
      Washington: National Academy of Engineering, 1974.
      141pp. Illus. $5.75.

102   O'Connor, Egan. "Moratorium Politics." Not Man
      Apart 3(5): 10-12. Drawings, table.

103   Odum, Howard T. Environment, Power, & Society.
      New York: Wiley-Interscience, 1971. Index, refs.

104   Orbis Scientiae, University of Miami, 1974. Topics in
      Energy and Resources. New York: Plenum, 1974.
      161pp. (Studies in the Natural Sciences, v. 7). Refs.
      $19.50.

105   Osann, Edward R. "In Search of New Solutions to the
      'Energy Crisis.'" Outdoor America 38(4): 8, April
      1973. Photos.

106   Othmer, Donad F. "Energy: Prospects for the Rest of
      the Century." Mechanical Engineering 96(8): 18-25.
      Drawings.

107   _____, and Roels, Oswald N. "Power, Fresh Water,
      and Food from Cold, Deep Sea Water." Science 182
      (4108): 121ff (5pp.), Oct. 12, 1973.

108   Pickering, William H. "Important to Examine Possibil-

ities of Unconventional Energy Sources. " Aware (31): 5-7, April 1973.

109 Pringle, Laurence P. Energy: Power for People. New York: Macmillan, 1975. Refs. $5.95. Juv. (Introduction to the earth's energy sources and to promising new alternatives to traditional sources of energy power.)

110 "Proposals Aim to Solve Energy Shortage. " Chemical and Engineering News 50(15): 2 April 10, 1972. Photo.

111 Rabinowitch, Eugene. "Challenges of the Scientific Age. " Bulletin of the Atomic Scientists 29(7): 4-8, Sept. 1973.

112 Radin, Alex. "How We Can Conserve Energy. " Public Power 31(4): 6-9, July-Aug. 1973.

113 "Rattlesnakes Are Harmless--Unless, Of Course, They Bite. " Not Man Apart 2(8): 1, Aug. 1972. (Editorial), drawing.

114 "Research for Power. " (editorial) New York Times. p. 40.

115 Ridgeway, James and Conner, Bettina. New Energy: Understanding the Crisis and a Guide to an Alternative Energy System. Boston: Beacon Press, 1975. Index. $7.95.

116 Risser, Hubert E. "The U.S. Energy Dilemma: The Gap Between Today's Requirements and Tommorrow's Potential. " Illinois State Geological Survey Report 64, July 1973. 59pp. Diagrs., graphs, tables.

117 Roberts, Ralph. "Energy Sources and Conversion Techniques. " American Scientist 61(1): 66ff (10pp.), Jan.-Feb. 1973. Diagrs., graphs, photo., refs., tables.

118 Roddis, Louis H., Jr. "Unconventional Power Sources." A speech before Chief Executives Conference of Edison Electric Institute, Dec. 4, 1970.

119 Rothman, Milton A. Energy and the Future. New
     York: Franklin Watts, 1975. 114pp. Illus. , refs.
     $5.90 PLB Juv.

120 Scheffer, Walter F. , ed. Energy Impacts on Public
     Policy and Administration: (papers). Conference on
     Energy and Its Impact on Public Policy and Adminis-
     tration, University of Oklahoma, 1974. Norman,
     OK: Advanced Programs, University of Oklahoma,
     1974. 238pp. Refs. $2.50 (pap).

121 Seaborg, Glenn T. , et al. "Nuclear and Other Energy."
     New York Academy of Sciences Annals 216: 79ff
     (10pp.), May 18, 1973. Table.

122 "Senate Approves Energy Reorganization Bill. " Con-
     gressional Quarterly 32(34): 2323-25, Aug. 24, 1974.

123 Shachtman, Tom. "Getting More Power to the People."
     Ecology Today 1(7): 46-50, Sept. 1971. Photos.

124 Shepherd, Jack. "Energy 7: The Alternatives. " Intel-
     lectual Digest 3: 21+, June 1973.

125 Shoupp, W. C. "Technological Possibilities for Future
     Power. " Aware (10): 3-6, April 1972.

126 Shuttlesworth, Dorothy. Disappearing Energy. Garden
     City, NY: Doubleday, 1974.

127 Sillin, Lelan F. , Jr. "A 30-year Plan for Pollution-
     Free Energy Sources. " New York Times, p. 39,
     Dec. 13, 1971.

128 Simon, Andrew. Energy Resources. Elmsford, NY:
     Pergamon, 1974.

129 Smith, Gene. "U. S. Sees Its Role in Energy Growing."
     New York Times p. 41, Sept. 25, 1971.

130 Snyder, M. Jack and Chilton, Cecil. "Planning for Un-
     certainty: Energy in the Years 1975-2000. " Battelle
     Research Outlook 4(1): 2-6, 1972. Graphs.

131 "Some Aspects of R and D Strategy in the United
     States. " Electrical Review 193(7): 222-26, Aug. 17,
     1973. Diagrs. photos.

132 Starr, Chauncy. "Research On Future Energy Sys-
     tems. " Professional Engineer 42(2): 20-25, Feb.
     1972. Chart, Drawings.

133 Steadman, Philip. Energy, Environment and Building.
     New York: Cambridge University Press, 1975.
     $15. 00. $5. 95 (pap).

134 Steinhart, Carol E. and Steinhart, John. The Fires of
     Culture: Energy Yesterday and Tomorrow. North
     Scituate, MA: Duxbury Press, 1974. 273pp. Illus.
     refs., index. $4. 95 (pap).

135 Stoner, Carol Hupping, ed. Producing Your Own
     Power: How To Make Nature's Energy Sources Work
     for You. Emmaus, PA: Rodale Press, Inc. 1974.
     Illus., refs., glossary, conversion tables, index.
     $8. 95. New York: Vintage Books, 1974. $2. 45
     (pap).

136 Summers, Claude M. "The Conversion of Energy. "
     Scientific American 224(3): 148ff (13pp.), Sept. 1971.
     Charts, diagr., photo.

137 Szego, C. G. "The U. S. Energy Problem, Vol. II:
     Appendices-Part A. " NTIS Report PB-207 518.
     744pp., Nov. 1972. Diagrs., graphs, tables.

138 Szulc, Tad. The Energy Crisis. New York: Watts,
     1974. 160pp. Refs., index. Gr. 9-12. $5. 95 PLB.

139 Thring, M. W. Society for Social Responsibility in
     Science Combustion 44(11): 19-21, May 1973.

140 Tilton, John E. U. S. Energy R & D Policy: The Role
     of Economics. Washington: Resources for the
     Future, 1974. 134pp. RFF Working Paper. EN-4.
     Refs. $3. 50 (pap).

141 A Time to Choose: America's Energy Future. (Energy
     Policy Project of the Ford Foundation.) Cambridge,
     MA: Ballinger, 1975. 512pp. $12. 95. $4. 95 (pap).

142 Udall, Stewart L. Beyond the Energy Crisis. McGraw
     Hill, 1974.

143 _____. The Energy Balloon. New York: McGraw-
     Hill, 1974. $7. 95.

144 _____ . "The Energy Crisis: A Radical Solution. "
World 2(10): 34-36, May 8, 1973. Photo.

145 United Nations. Proceedings of the United Nations Conference on New Sources of Energy: Solar Energy,
Wind Power, and Geothermal Energy, Rome, 21-31
Aug. 1961. New York: United Nations, 1964.
$33.50 for set of 7 vols. Vol. 1 General Sessions,
218pp. $2.50; Vol. 4 Solar Energy I, 665pp., illus.,
$7.60, electricity; Vol. 5 Solar Energy II, 423pp.,
illus., $4.50, heating; Vol. 6 Solar Energy III,
Cooling, distillation, furnaces.

146 U.S. Congress. House of Representatives. Committee
on Science and Astronautics. "Energy Research and
Development. " 92nd Congress LL Serial EE, Washington, Dec. 1972. 418pp. Diagrs., graphs, tables.

147 "Utilization of Energy. " Mosaic 5(2): 33-37, Spring
1974.

148 "Wanted: A Coordinated, Coherent National Energy
Policy Geared to the Public Interest. " Conservation
Foundation Letter, June 1972. 12pp.

149 Ware, Warren W. World Energy Supply and Demand.
Cleveland, OH: Predicasts, Inc. $375.00.

150 Weaver, Kenneth F. "The Search for Tomorrow's
Power. " National Geographic 142(5): 650-82, Nov.
1972. Photos., diagrs.

151 Weinberg, Alvin M. "Long-Range Approaches for Re-
solving the Energy Crisis. " Mechanical Engineering 95(6): 14ff (5pp.), June 1973. Diagrs., photos.

152 Weingarten, Jerome. "Surviving the Energy Crunch. "
Environmental Quality Magazine 4(1): 28ff (7pp.),
Jan. 1973. Photos., table.

153 Wolozin, Harold, ed. Energy and the Environment: Selected Readings. Morristown, NJ: General Learning
Corp., 1974. 133pp. Refs. $3.10 (pap).

154 Yannacone, Victor J. Energy Crisis: Danger and Op-
portunity. St. Paul, MN: West Pub. Co., 1974.
432pp. Refs.

2

SOLAR ENERGY

1000 Abetti, Giorgio. Solar Research. New York: Mac-
     millan, 1962. 173pp. Photos., figures, index.

1001 Abbott, C. G. "Solar Power from Collecting Mirrors."
     Solar Energy Research. Madison: University of
     Wisconsin Press, 1961.

1002 Abou-Hussein, M. S. M. "Temperature-Decay Curves
     in the Box Type Solar Cooker." United Nations
     Conference on New Sources of Energy, E35-S75,
     Rome, 1961. New York: United Nations, 1964.

1003 Achenbach, P. R. and Cable, J. B. "Site Analysis
     for the Applications of Total Energy Systems to
     Housing Developments." 7th Intersociety Energy
     Conversion Engineering Conference, San Diego, CA.
     Sept. 1972.

1004 Achilov, B., et al. "Investigation of an Industrial-
     Type Solar Still." Geliotekhnika 7(2): 33-36, 1971.
     (Russian).

1005 Adams, W. C. and Day, R. E. "The Actions of Light
     on Selenium." Proceedings, Royal Society 25: 113-
     17, London, 1876.

1006 "Africa Plugs Into the Sun." UNESCO Courier 27(1):
     30-31, Jan. 1974. Photos.

1007 "AIA Presents Views on Solar Energy: Need for Archi-
     tect Involvement is Stressed." MEMO, March 5,
     1974.

1008 Akyurt, M. and Selcuk, M. K. "A Solar Drier Supple-
     mented with Auxiliary Heating Systems for Contin-
     uous Operation." Solar Energy 14: 313-20, 1973.

1009 Alfven, Hannes. "Energy and Environment." Bulletin of the Atomic Scientists 18(5): 5-8, May 1972.

1010 Allegretto, M. "Power from the Sun: A New Look at an Old Idea." National Parks and Conservation Magazine 48(1): 13-15, Jan. 1974. Illus., diagr., drawing.

1011 "Alternative Energy Sources May Help Stem the Energy Crisis." Product Engineering 44(10):22-26, Oct. 1973. Illus., diagrs.

1012 "The Alternatives, Research and Development (R & D) for New Energy Sources." Economic Priorities Report 3(2): 23-25, May-June 1972.

1013 Altman, Manfred. "Conservation and Better Utilization of Electric Power by Means of Thermal Energy Storage and Solar Heating." NTIS Report PB-210 359, Oct. 1971. 263pp.

1014 Anderson, B. Solar Energy and Shelter Design. Available from author, Box 47, Harrisville, NH. 15pp.

1015 Anderson, D. B., Erickson, G. A., Jordan, R. C., and Leonard, R. R. "Field Laboratory for Heating Studies." American Society of Heating, Refrigerating, and Air-Conditioning Engineers, Nov. 1960.

1016 Anderson, J. H. "The Sea Plant--A Source of Power, Water, and Food Without Pollution." Solar Energy 14: 287-300, 1973.

1017 _____, and Anderson, J. H., Jr. "Thermal Power from Sea-Water." Mechanical Engineering 88(4): 41-46, April 1966.

1018 Anderson, L. B., Hottel, H. C., and Whillier, A. "Solar Heating Design Problems." Solar Energy Research. Madison: University of Wisconsin Press, 1955. (pp. 47-56)

1019 _____, Hunter, J. M., Dietz, R. H., and Close, W. A. "The Architectural Problems of Solar Collectors: A Round Table Discussion." Proceedings, World Symposium on Applied Solar Energy, Phoenix, Arizona, 1955, Menlo Park, CA: Stanford Research Institute, 1956.

1020   Anderson, Richard J.   "The Promise of Unconven-
        tional Energy Resources. " Battelle Research Out-
        look 4(1): 22ff (4pp.), 1972.   Photos.

1021   Anderson, W. A. and Delahoy, A. E.   "Schottky Bar-
        rier Diodes for Solar Energy Conversion. " Pro-
        ceedings of the IEEE 60 (11): 1457-58, Nov. 1972.
        Illus. , diagr.

1021a   _____, et al.   "8% Efficient Layered Schottky-
        Barrier Solar Cell. "   Journal of Applied
        Physics 45(9): 3913-15, Sept. 1974.   Refs. , illus. ,
        diagrs.

1022   Andrassy, S.   "Solar Water Heaters. "   United Nations
        Conference on New Sources of Energy, E35-S96,
        Rome, 1961.   New York: United Nations, 1964.

1023   Angstrom, A.   "Solar and Terrestrial Radiation. "
        Journal of the Royal Meterological Society 50(121),
        1924.

1024   Annaev, A. , Bairamov, R. , and Rybakova, L. E.
        "Effect of Wind Speed and Direction on the Output
        of a Solar Still. " Geliotekhnika 7(4): 33-37, 1971
        (in Russian).

1025   Applications Thermique de l'Energie Solaire dans le
        Domaine de la Recherche et de l'Industrie, Sym-
        posium at Mont-Louis, France, 1958.   Paris:
        Centre National de la Recherche Scientifique, 1961.
        (Articles by F. Trombe, 87-129; A. Le-Phat-
        Vink, 145-62; V. Baum, R. Aparissi, and D.
        Tepliakoff, 163-74; F. Trombe, M. Foex, and
        C. H. LaBlanchetais, 174-214; P. Glazer, 215-34;
        F. Trombe and M. Foes, 277-319; E. Roger, 319-
        26; M. Foex, 327-66; C. Rekar, 367-82; F.
        Trombe and M. Foex, 423-50.)

1026   "Archimedes As Arsonist. " Newsweek 82(22): 64,
        Nov. 26, 1973. Illus.

1027   "Archimedes' Weapon: Explanation to Use of Solar
        Power to Set Fire to Roman Ships. " Time 102
        (22): 60, Nov. 26, 1973. Illus.

1028   "Architettura Solare. " Domus (528): 5-8, Nov. 1973.
        Illus. , plans, diagrs.

1029   Arnold, W. F.   "Will Solar Cells Shine on Earth?"
        Electronics 45(11): 67-69, May 22, 1972.   Photo,
        drawing.

1030   Arthur D. Little, Inc.   "Industry Moves Toward a
        Solar Climate Control Market."   Professional Engi-
        neer 43(10): 26ff (4pp.), Oct. 1973.   Drawing.

1031   Armour Research Foundation.   "Investigation of Solar
        Energy for Ice Melting."   Final Report June 1,
        1949.   Chicago: Armour Research Foundation, 1949.

1032   Ashar, N. G. and Reti, A. R.   "Engineering & Eco-
        nomic Study of the Use of Solar Energy Especially
        for Space Cooling in India and Pakistan."   United
        Nations Conference on New Sources of Energy,
        E35-S37, Rome, 1961.   New York: United Nations,
        1964.

1033   ASHRAE Handbook of Fundamentals.   New York:
        American Society of Heating, Refrigerating and Air
        Conditioning Engineers, 1972.

1034   An Assessment of Solar Energy as a National Energy
        Resource.   Washington, DC: U. S. Superintendent
        of Documents, 3800-00164.   85pp.

1035   Association for Applied Solar Energy.   Living with the
        Sun.   Phoenix, AZ: Association for Applied Solar
        Energy, 1958.   Illus., maps, diagrs., plans.

1036   _____ .   Proceedings of the 1957 Solar Furnace
        Symposium.   Phoenix, AZ: Association for Applied
        Solar Energy, 1957.   116pp.   Illus., charts, diagrs.

1037   _____ .   Proceedings of the World Symposium on
        Applied Solar Energy, 1955.   Menlo Park, CA:
        Stanford Research Institute, 1955.   Avail. also
        New York: Johnson Reprint Corp., 1955.   Illus.,
        portraits. $15.00.

1038   "Audubon Society Building Will Employ Solar System."
        Air Conditioning, Heating, and Refrigeration News
        131(2): 23, Jan. 14, 1974.   Diagr.

1039   "Australia Has Practical Experience of Solar Heating."
        Energy Digest 2(5): 8-10, Sept.-Oct. 1973.   Photo.

1040 "The Available Energy, The Measurement of Radiation." Transactions of the Conference on the Use of Solar Energy: The Scientific Basis, Vol. 1. Tucson: University of Arizona Press, 1958.

1041 Axtmann, Robert C. "Padre Sol and Chemical Fuel Production." Bulletin of the Atomic Scientists 29 (8): 42-45, Oct. 1973. Diagrs.

1042 Ayers, D. L. "Temperature Predictions During Spacecraft Maneuvers." Journal of Spacecraft and Rockets 6: 1343-44, Nov. 1969. Diagrs.

1043 Ayres, E. and Scarlott, C. A. Energy Sources, the Wealth of the World. New York: McGraw-Hill, 1952.

1044 Backus, C. E. "A Solar-Electric Residential Power System." Proceedings of the Intersociety Energy Conversion Engineering Conference (IECEC), 1972.

1045 Bacon, F. T. "Energy Storage Based on Electro-lyzers and Hydrogen-Oxygen Fuel Cells." United Nations Conference on New Sources of Energy, E35-Gen. 9, Rome, 1961. New York: United Nations, 1964.

1046 Bailey, R. L. "Proposed New Concept for a Solar-energy Converter." Journal of Engineering for Power 94: 73-77, April 1972. Refs., diagrs., illus.

1047 "Balloon Batteries, Charged and Heated by Solar Energy." Mechanical Engineering 92: 50, April 1970. Diagrs.

1047a Barnes, Peter. "Solar Derby, Who'll Control Sun Power?" New Republic 172(5): 17-19, Feb. 1, 1975.

1048 Bartz, D. R. and Horsewood, J. L. "Characteristics, Capabilities, and Costs of Solar Electric Spacecraft for Planetary Missions." Journal of Spacecraft and Rockets 7: 1379-90, Dec. 1970. Refs., illus., diagrs.

1049 Bassler, Friedrich. "Solar Depression Power Plant

of Qattara in Egypt. " Solar Energy 14(1): 21ff
(8pp.), Dec. 1972. Graphs, map.

1050 Baum, V. A. "Prospects for the Application of Solar
Energy and Some Research Results in the U. S. S. R. "
Proceedings, World Symposium Applied Solar En-
ergy, Phoenix, Arizona, 1955. Menlo Park, CA:
Stamford Research Institute, 1956. pp. 289-98.

1051 _____ . Semiconductor Solar Energy Converters.
New York: Plenum Press, 1969. $27. 50.

1052 _____ . "Solar Distillers. " United Nations Conference on New Sources of Energy, E35-S119, Rome,
1961. New York: United Nations, 1964.

1053 _____ . "The Use of Solar Energy for Electricity
Production by Direct Conversion by Means of
Thermoelectric Converters & Photoelectric Cells. "
United Nations Conference on New Sources of En-
ergy, E35-Gr-S10, Rome, 1961. New York: United
Nations, 1964.

1054 _____ , Aparase, A. R. , and Garf, B. A. "High-
Power Solar Installations. " Solar Energy 1(2): 6-
13, 1957.

1055 _____ , and Strong, J. D. "Basic Optical Considerations in the Choice of a Design for a Solar
Furnace. " Solar Energy 2(3-4): 37-45, 1958.

1056 Bayless, J. R. , et al. "Feasibility of High Voltage
Solar Arrays. " Journal of Spacecraft and Rockets
9: 522-28, July 1972. Refs. , diagrs.

1057 Beckman, W. A. , Duffie, J. A. , and Lof, G. O. G.
Literature Review of Solar Heating and Air Condi-
tioning. Madison: University of Wisconsin, Solar
Energy Laboratory, 1971. 8pp.

1058 _____ , Schaffer, P. , Hartman, W. R. , Jr. , and
Lof, G. O. G. "Design Consideration for a 50 Watt
Photovoltaic Power System Using Concentrated
Solar Energy. " Solar Energy 10(3): 132-36, 1966.

1059 Behrman, Dan. "Solar Energy Claims a New Place
In the Sun. " UNESCO Courier 27(1): 24-27, Jan.
1974. Photo.

1060 Belanger, J. D. "How to Use the Sun Around the Home." Organic Gardening and Farming 19(1): 74-6, Jan. 1972. Drawing, plan.

1061 Belton, G. and Ajami, F. "Thermochemistry of Salt Hydrates." Report NSF/RANN/SE/GI27979 TR/73/4 of the University of Pennsylvania National Center for Energy Management and Power to NSF, 1973.

1062 Benford, F. and Bock, J. E. "A Time Analysis of Sunshine." Transactions of American Illuminating Engineering Society 34: 200+, 1939.

1063 Benveniste, Guy. Solar Energy Furnaces: A Review of Development in the Utilization of Solar Energy to Produce High Temperatures. Menlo Park: Stanford Research Institute, 1955. 23pp.

1064 Berks, W. I. and Luft, W. "Photovoltaic Solar Arrays for Communications Satellites." IEEE Proceedings 59(2): 263-71, Feb. 1971. Graphs, photos, diagrs., refs.

1065 Berman, P. A. Development of Lithium-Doped Radiation Resistant Solar Cells. Pasadena, CA: Jet Propulsion Lab, 1972. 25pp. NTIS #N-73-11044 GA.

1066 _____. Photovoltaic Solar Array Technology Required for Three Wide Scale Generating Systems for Terrestrial Applications Rooftop, Solar Farm, and Satellite. Pasadena, CA: Jet Propulsion Lab, California Institute of Technology, 1972. 27pp. NTIS Accession #N72-33061.

1067 Bernard, J. and Mottet, S. Partial Results on CDS Solar Cells. (Trans. of Resultats Partiels sur les Cellules Solaires au CDS.) Toulouse: Office National Detudes et de Recherches Aerospatiales, 1972. 36pp. NTIS Accession #N73-12062.

1068 Bernard, L. "Production d'Air Chaud par le Dépositif de la Cheminée Solaire." Applications Thermiques de l'Energie Solaire dans le Domaine de la Recherche et de l'Industrie. Paris: Centre de la Recherche Scientifique, 1961. pp. 641-49.

1069 Bevill, V. D. and Brandt, H. "A Solar Energy Collector for Heating Air." Solar Energy 12: 19-29, 1968.

1070 Bhaskara Rao, A. and Padmanabhan, G. R. "Method for Estimating the Optimum Load Resistance of a Silicon Solar Cell Used in Terrestrial Power Applications." Solar Energy 15(2): 171-77, July 1973.

1071 "Billion Dollar Solar Energy Market Hinted by 1985." Architectural Record 154(2): 34, Aug. 1973. Diagr.

1072 Bird, R. B., Stewart, W. C., and Lightfoot, E. N. Transport Phenomena. New York: Wiley, 1960.

1073 Bischof, F. V. Space Station Solar Array Technology Evaluation Program (Final Report). Sunnyvale, CA: Lockheed, 1973. 138pp. NTIS #73-30057/6GA.

1074 Bjorksten, J. and Lappala, R. P. "Photodegradation of Plastic Films." Plastics Technology, Jan. 1957.

1075 Black, J. N. "Some Aspects of the Climatology of Solar Radiation." United Nations Conference on New Sources of Energy, E35-S13, Rome, 1961. New York: United Nations, 1964.

1075a Blackwell, G. R. "Solar Cells in the Future." Electronics and Power 20: 707-08, Sept. 19, 1974.

1076 Blanco, P. "Solar Energy Availability and Instruments for Measurements." United Nations Conference on New Sources of Energy, E35-GrS11, Rome, 1961. New York: United Nations, 1964.

1077 Bleksley, A. E. H. "The Intensity of Solar Radiation in Southern Africa." Transactions of the Conference on the Use of Solar Energy: The Scientific Basis, Vol. 1: 132-35, 1958. Tucson: University of Arizona Press, 1958.

1078 Bliss, R. W. "Atmospheric Radiation Near the Surface of the Ground." Solar Energy 5: 103+, 1961.

1079 _____. "The Derivations of Several Plate Efficiency Factors." Solar Energy 3(4), Dec. 1959.

22 / Alternate Sources of Energy

1080 _____. "Design and Performance of the Nation's
Only Fully Solar-Heated House." Air Conditioning,
Heating and Ventilating 92, Oct. 1955. See also
Proceedings of the World Symposium on Applied
Solar Energy 151, SRI, Menlo Park, CA, 1956.

1081 _____. "The Performance of an Experimental System Using Solar Energy for Heating & Night Radiation for Cooling a Building." United Nations Conference on New Sources of Energy, E35-S30, Rome,
1961. New York: United Nations, 1964.

1082 Bloemer, J. W., Collins, R. A., and Eibling, J. A.
"Field Evaluation of Solar Sea Water Stills." Ad-
vances in Chemistry, Series 27. Washington, DC:
American Chemical Society, 1960. (8)

1083 _____, _____, and _____. Study and Evaluation of Solar Sea Water Stills, PB 171, 934.
Washington, DC: Department of Commerce, Office
of Technical Services, 1961.

1084 _____, and Eibling, J. A. "A Progress Report on
Evaluation of Solar Sea Water Stills." American
Society of Mechanical Engineer Paper 61-WA-296,
1961.

1085 "Blunt Talk to Solar Energy Supporters." Science
News 105(5): 69-70, Feb. 2, 1974.

1086 Blytas, G. C., Kertesz, D. J., and Daniels, F.
"Concentrated Solutions in Liquid Ammonia; Vapor
Pressures of LiNo3-NH3 Solutions." Journal of the
American Chemical Society 84: 1083-85, 1962.

1087 Böer, Karl W. "A Combined Solar Thermal Electrical
House System." Paper presented at International
Solar Energy Society Conference, Paris, 1973.

1088 _____. "Future Large Scale Terrestrial Use of
Solar Energy." Proceedings of the NASA Conference on Solar and Chemical Power, QC-603, p. 145-
148, 1972.

1089 _____. "Solar Collectors of the University of Delaware Solar House Project." Proceedings of the
NSF/RANN Solar Heating and Cooling for Buildings

Workshop, NSF/RANN-73-004, pp. 15-16, July
1973.

1090 _____. "Solar Heating and Cooling of Buildings--
Results and Implications of the Delaware Experiment." Energy Environment Productivity, Proceedings of the First Symposium on Rann: Research
Applied to National Needs, Washington, DC, Nov.
18-20, 1973. Washington, DC: National Science
Foundation. p. 50-53. Photos., diagrs., graph.

1091 _____. "The Solar House and Its Portent." Chemtech ACS 3(7): 394ff (7pp.), July 1973. Drawing,
diagrs., graphs, photo.

1092 _____, Telkes, M., and Trela, J. A. "Solar One:
A Systems Approach to Solar Energy Conversion."
Naturalist, 1973.

1093 Boleszny, Ivan. Solar Heating. (Research Services
Biblio. Sers. 4 No. 88). Adelaide: State Library
of South Australia, 1967. 34pp. Refs.

1094 Bolin, J., Tunukest, C. J., and Milner, C. J.
"Plastic-Replica Mirror Segments for a Solar Furnace." Solar Energy 5(3): 99-102, 1961.

1095 Boretz, J. E. "Large Space Station Power Systems."
Journal of Spacecraft and Rockets 6: 929-36, Aug.
1969. Refs., diagrs. Discussion, W. G. Ruehle 7:
639; Reply 639-40, May 1970.

1096 Boyd, T. "Pool-Heating Panels Make the Sun Work
For You." Popular Science 202(5): 117-118, May
1973. Drawings, photos.

1097 Brabyn, H. "Well of Knowledge; Water Pump Powered
by Sunlight in Chinguetti, Mauritania." UNESCO
Courier 27(1): 28ff (3pp.), Jan. 1974. Photos.,
drawings.

1098 Brainin, Sally. "New Uses of an Old-Fashioned Energy
Source." Conservation News 38(7): 2-5, April 1,
1973. Photo.

1099 Brand, David. "Catching Sunbeams." Wall Street
Journal, p. 1, April 16, 1973.

1100 Branley, Franklyn Mansfield. Solar Energy. New
       York: Crowell, 1957. 117p. Illus. (Gr. 5-8)
       $3.95.

1101 "Breakaway House That's Heated by the Sun: Solar-
       Heated House Near Albuquerque." House & Garden
       145(1): 72-75, Jan. 1974. Photos., plan.

1102 Breihan, R. R., Daniels, F., Duffie, J. A., and
       Löf, G. O. G. "Preliminary Tests of a Solar-
       Heated Thermoelectric Converter." United Nations
       Conference on New Sources of Energy, E35-S103,
       Rome, 1961. New York: United Nations, 1964.

1103 Bridgers, Frank H. "Applying Solar Energy for Cool-
       ing and Heating Institutional Buildings." Papers
       presented at the Symposium on Solar Energy Appli-
       cations at the ASHRAE Annual Meeting, June 23-27,
       1974, Montreal, Canada. ASHRAE Journal 16
       (9): 29-37, Sept. 1974. Diagrs., illus., maps,
       flow diagrs.

1104 _____. "Energy Conservation: Pay Now, Save
       Later." ASHRAE Journal, Oct. 1973. p. 47.

1105 Briefings Before the Task Force on Energy, Vol. 3.
       Report Hs Comm Science Astronautics 92 Con II
       Serial U, May 1972. 204pp.

1106 Brimelow, T. Solar Energy Technology: 1954-1959.
       London: Library Association, 1961. 32pp. Refs.

1107 Brinkworth, Brian Joseph. "Prospects for Solar
       Power." Energy Digest 2(5): 2ff (4pp.), Sept. -
       Oct. 1973. Photo., table.

1108 _____. "Solar Energy." Nature 249(5459): 726ff
       (3-1/3pp.), June 21, 1974.

1109 _____. Solar Energy For Man. Salisbury: Comp-
       ton Press Ltd., 1972. 250pp. Illus., index,
       refs. (In U.S., New York: J. Wiley, 1973. 251pp.
       $8.95.

1110 Brooks, F. A. "Notes on Spectral Quality and Mea-
       surement of Solar Radiation." Solar Energy Re-
       search. Madison: University of Wisconsin Press,
       1955. pp. 19-29.

1111 _____. Use of Solar Energy for Heating Water. (Publ. 3557). Washington, DC: Smithsonian Institution, 1939.

1112 Brown, William C. "Satellite Power Stations: A New Source of Energy?" IEEE Spectrum 10(3): 38-47, March 1973. Diagrs., Drawing, graphs, refs., tables.

1113 Brunt, D. "Notes on Radiation in the Atmosphere." Journal of the Royal Meterological Society 58: 389-420, 1932.

1114 Buch, J. M. "Clean Power: Amateur Radio's Challenge for the Future." QST 56(8): 64-67, Aug. 1972. Refs., photos., diagrs.

1115 Buchberg, H., et al. "Performance Characteristics of Rectangular Honeycomb Solar Thermal Converters." Solar Energy 13(2): 193-217, May 1971.

1116 _____, and Roulet, J. R. "Simulation and Optimization of Solar Collection and Storage for House Heating." Solar Energy 12(2): 31, Sept. 1968.

1117 Buelow, F. H. "Drying Crops with Solar Heated Air." United Nations Conference on New Sources of Energy, E35-S17, Rome, 1961. New York: United Nations, 1964.

1117a "Building Owners Research, Retrofit to Use Sun." Engineering News-Record 192(12): 21, Sept. 12, 1974.

1118 "Building's Energy Design Uses Soil, Sun, and Plants." Engineering News-Record 192(8): 15, Feb. 21, 1974.

1119 Burda, E. J., ed. Applied Solar Energy Research: A Directory of World Activities and Bibliography of Significant Literature. Menlo Park, CA: Stanford Research Institute, 1955. 289pp.

1120 Burke, J. C. "Massachusetts Audubon Society Solar Building." Proceedings of the NSR/RANN Heating and Cooling for Buildings Workshop, NSF/RANN-73-004, pp. 214-17, July 1973.

1121 Burlew, J. S. Algal Culture from Laboratory to Pilot
      Plant, Publ. 600. Washington, DC: Carnegie In-
      stitute of Washington, 1953.

1122 Burns, Raymond K. Preliminary Thermal Performance
      Analysis of the Solar Brayton Heat Receiver. Wash-
      ington, DC: NASA sale through Clearinghouse for
      Federal Scientific and Technical Information, Spring-
      field, VA, 1971. 40pp. Illus. (NASA TN D-6268)
      70-611885.

1123 Busch, C. W. Future Portable Energy Systems.
      (M. S. Thesis.) Madison: University of Wisconsin
      Press, 1962.

1124 Bush, Clinton G., Jr. "'Supercore' Eliminates House-
      hold Waste and Pollution at the Source." Catalyst
      for Environmental Quality 3(2): 22-26, 1973.
      Diagrs.

1125 Butler, C. P., et al. "Surfaces for Solar Spacecraft
      Power." Solar Energy 8(1): 2-8, Jan.-March 1964.

1126 Butz, L. W. "Use of Solar Energy for Residential
      Heating and Cooling." M. S. Thesis in Mechanical
      Engineering, Madison: University of Wisconsin,
      1973.

1127 Calvert, J. G. "Photochemical Processes for Utili-
      zation of Solar Energy." Introduction to the Utili-
      zation of Solar Energy. (Zarem, A. M. and Enway,
      D. D., eds.)

1128 Calvin, M., and Brassham, J. A. The Path of Carbon
      in Photosynthesis. Englewood Cliffs, NJ: Pren-
      tice-Hall, 1957.

1129 _____, and _____. The Photosynthesis of Carbon
      Compounds. New York: Benjamin, 1962.

1130 Carpenter, E. K. "Doggone It, Our Heating Fuel-Oil
      Bill Is Up to $6 This Month." American Home
      77(11): 20, Sept. 1974.

1131 Caryl, C. R. "Ratio of Ultraviolet Total Radiation in
      Phoenix." Proceedings of World Symposium on
      Applied Energy, Phoenix, 1955. Menlo Park, CA:

Stanford Research Institute, 1956.

1132   "A Case for a Solar Ice Maker." Solar Energy 7(1):
       1-2, 1963.

1133   Cellarius, Richard. "Are Solar Power Systems Really
       Pollution Free?" Alternative Sources of Energy
       (11): 8-10.

1134   "Chance for Solar Energy Conversion." Chemical &
       Engineering News 49(52): 39, Dec. 20, 1971.

1135   Chang, Sheldon S. L. Energy Conversion. Englewood
       Cliffs, NJ: Prentice-Hall, 1963. 237pp.

1136   Chapin, D. M. "The Direct Conversion of Solar Energy
       to Electrical Energy." Introduction to the Utiliza-
       tion of Solar Energy. (Zarem, A. M. and Enway,
       D. D., eds.) New York: McGraw-Hill, 1963, pp.
       153-89.

1137   Charters, W. W. S. and Peterson, L. F. "Free Con-
       vection Suppression Using Honeycomb Cellular Ma-
       terials." Solar Energy 13(4): 353-61, July, 1972.

1138   "Chauffage Solaire." L'Architecture d'Aujourd'hui
       (156): viii, June 1971. Illus., plans, diagr.

1139   Cheban, A. G., et al. "The P-Type-GAAS (1-x)PX
       N-Type GAAS Heterojunction Solar Cells." Gelio-
       tekhnika 27(1): 30-33, 1972. Trans. by Transla-
       tions Consultants Ltd. NTIS #72-32075.

1140   Chedd, Graham. "Brighter Outlook for Solar Power."
       New Scientist 58(840): 36-38, April 5, 1973. Draw-
       ing.

1141   Cherry, William R. "The Generation of Pollution Free
       Electrical Power from Solar Energy." Research
       Report, NASA, Goddard Space Flight Center, March
       1971. 15pp. Charts, diagrs., photos. Also in
       Journal of Engineering for Power 94: 78-82, April
       1972. Illus., map, diagrs.

1142   _____. "Harnessing Solar Energy: The Potential."
       Astronautics & Aeronautics 11(8): 30-37, Aug. 1973.
       Diagr., graphs, table, drawing, refs.

1143 _____. Proceedings 13th Annual Power Sources Conference, Atlantic City, NJ, pp. 62-66, May 1959.

1144 Chinnappa, J. C. V. "Computed Year-Round Performance of Solar-Operated Multi-State Vapour Absorption Air Conditioners at Georgetown, Guyana, and Colombo, Ceylon." Paper presented at International Solar Energy Congress, Paris, 1973.

1145 _____. "Experimental Study of the Intermittent Vapor Absorptions Refrigeration Cycle Employing the Refrigerant-Absorbent Systems of Ammonia-Water and Ammonium-lithium-nitrate." Solar Energy 5: 1-8, 1961.

1146 _____. "Performance of an Intermittent Refrigerator Operated by a Flat-Plate Collector." Solar Energy 6: 143-150, 1962.

1147 _____. "The Solar Operation of a Vapour Absorption Cycle Air Conditioner at Colombo." Trasn. Inst. Engrs., Ceylon 19, 1967.

1148 Chiou, J. J., El-Wakil, M. M., and Duffie, J. A. "A Slit and Expanded Aluminum Foil Solar Collector." Solar Energy Symposium, University of Florida, April 1964; Ph.D. Thesis, J. J. Chiou, University of Wisconsin, 1964.

1149 Choudhury, N. K. D. "Solar Radiation at New Delhi." Solar Energy 7: 44-52, 1963.

1150 Chung, R. Letter to the editor. Solar Energy 7: 1-2, 1963.

1151 _____, and Duffie, J. A. "Cooling with Solar Energy." United Nations Conference on New Sources of Energy, E35-S82, Rome, 1961. New York: United Nations, 1964.

1152 _____, Löf, G. O. G., and Duffie, J. A. "Experimental Study of a LiBr-H2O Absorption Air Conditioner for Solar Operation." American Society of Mechanical Engineers, Paper 62-WA-347, 1962.

1153 _____, _____, and _____. "A Study of a Solar Air Conditioner." Mechanical Engineering 85: 31, 1963.

1154 Clark, Wilson. "Beyond the Energy Crisis." Not Man Apart 1(11): 8-9, Nov. 1971. Photos.

1155 _____. Energy for Survival: Alternative to Extinction. Garden City, NY: Anchor Press/Doubleday. Illus. $12.50.

1156 _____, et al. "How to Kick the Fossil Fuel Habit." Environmental Action 5(18): 3-6, 12-15, Feb. 2, 1974. Drawings.

1157 "Clean Energy." Message from the President of the United States, June 4, 1971. 12pp.

1158 Close, D. J. "Flat-Plate Solar Absorbers: The Production and Testing of a Selective Surface for Copper Absorber Plates." Report E.D. 7, Engineering Section, Commonwealth Scientific and Industrial Research Organization, Melbourne, Australia, 1962.

1159 _____. "The Performance of Solar Water Heaters with Natural Circulation." Solar Energy 6(1): 33, 1962.

1160 _____. "Rock Pile Thermal Storage for Comfort Air Conditioning." Mech. and Chem. Engr. Trans. Instit. Engrs., Australia MC1: 11, 1965.

1161 _____. "Solar Air Heaters." Solar Energy 7: 117-124, 1963.

1162 _____, Dunkle, R. V., and Robeson, K. A. "Design and Performance of a Thermal Storage Air Conditioner System." Mech. Chem. Engr. Trans., Inst. Engrs. Australia MC4: 45, 1968.

1162a Cloud, F. M. "ASHRAE Journal Article Leads to Georgia Solar Water Heater Experiment." ASHRAE Journal 16(9): 45-46, Sept. 1974. Diagr.

1163 CNES (Centre National d'Etudes Spatiales). Solar Cells. New York: Gordon & Breach Science Publisher, Inc., 1971. Illus. $29.50.

1164 Cobble, M. H. "Analysis of a Conical Solar Concentrator." Solar Energy 7: 75, 1963.

1165 _____. "Theoretical Concentrations for Solar Furnaces." Solar Energy 5(2): 61-72, 1961.

1166 Coblentz, W. W. "Harnessing Heat from the Sun." Scientific American 127(5): 324, Nov. 1922.

1167 Cohen, R. K. and Hiester, N. K. "A Survey of Solar Furnaces in the United States." Solar Energy 1 (2-3): 115-17, 1957. Also in Solar Energy 1(4): 35+, 1957.

1167a "Collecting Solar Energy by Chemical Reaction." Chemistry 47(10): 26-27, Nov. 1974. Diagr.

1168 Colorado State University--Westinghouse Report to NSF, Report NSF/RANN/SE/GI 37815/PR/73/3. "Solar Thermal Electric Power Systems."

1169 "Conference Center Will Use Solar Heat." Engineering News-Record 192(4): 22-23, Jan. 24, 1974. Photo.

1170 Conference of the Association for Applied Solar Energy, New York, 1959. "Solar Furnaces." Solar Energy 3(3): 39-48, 1959.

1171 Conference on the Use of Solar Energy: The Scientific Basis, Vol. 5, (Transactions) University of Arizona, 1955. Tuscon: University of Arizona Press, 1958. Illus., maps.

1172 Conference Record. 8th Photovoltaic Specialists Conference, Seattle, 1970. New York: Institute of Electrical and Electronic Engineers, 1970. 379pp. Illus.

1172a "Congress and Solar Energy: Legislative Summary." Architectural Record 157(1): 35, Jan. 1975.

1173 Conn, W. M. "Solar Furnaces for Attaining High Temperatures." Solar Energy Research. (Daniels, F. and Duffie, J. A., eds.). Madison: University of Wisconsin Press, 1955. pp. 163-67.

1174 _____. "The Importance of Accurate Temperature Measurements in Work with Solar Furnaces." Transactions of the Conference on the Use of Solar Energy: The Scientific Basis 2: 205-12. Tucson:

University of Arizona Press, 1958.

1175 "Contracts Let for Solar Heating. " Aviation Week & Space Technology 100(4): 44, Jan. 28, 1974.

1176 Cook, Earl. "Energy for Millennium Three. " Aware (20): 3-7, Feb. 1973.

1177 Cooper, P. I. "The Absorption of Solar Radiation in Solar Stills. " Solar Energy 12: 3+, 1969.

1178 _____. "Digital Simulation of Experimental Solar Still Data. " Solar Energy 14: 451, 1973.

1179 Copeland, A. W., Black, O. D. and Garrett, A. B. Chemistry Review 31: 177-226, 1942.

1180 Cotton, E., et al. "Image and Use of the U. S. Army Quartermaster Solar Furnace. " U. N. Conference on New Sources of Energy E35-S, Rome, 1961. New York: United Nations, 1964.

1181 Covault, C. "Energy Relay Satellites Urged for Shuttle. " Aviation Week & Space Technology 98(2): 47-48, Jan. 8, 1973. Photo.

1181a Cranston, Alan. "Bright Future for Solar Energy. " National Parks and Conservation Magazine 48(10): 10-13, Oct. 1974. Photos., diagrs.

1182 Crausse, E. and Gachan, H. "The Study of a Saharan Solar House. " United Nations Conference on New Sources of Energy, E35-S76, Rome, 1961. New York: United Nations, 1964.

1183 "The Crisis of Converting Fuel into Energy. " Business Week (2210): 58-60, Jan. 8, 1972. Charts.

1184 Cummerow, R. L. "The Theory of Energy Conversion in P-N Junctions. " Transactions of the Conference on the Use of Solar Energy: The Scientific Basis 5: 57-64. Tucson: University of Arizona Press, 1958.

1185 Currin, C. G., Ling, K. S., Ralph, E. L., Smith, W. A., and Stirn, J. R. "Feasibility of Low Cost Silicon Solar Cells. " Proceedings of the 9th IEEE

Photovoltaic Specialists Conference, Silver Spring, MD, May 2-4, 1972.

1186 "Curve to Capture the Sun: Odeillo in the Pyrenees." Life 66(20): 80-2, May 23, 1969. Photo.

1187 Cussen, W. G. "Solar Cells in Satellite Power Supplies." Electronics and Power 18: 93-95, March 1972. Illus., diagrs.

1188 Czarnecki, J. T. "A Method of Heating Swimming Pools by Solar Energy." Solar Energy 7: 3-7, 1963.

1189 Daniels, Farrington. "Atomic and Solar Energy." American Scientist 38: 521-48, 1950.

1190 _____. "Construction et Essais de Reflecteurs Solaires Focalisants sans Verre." Applications Thermiques de l'Energie Solaire dans le Domaine de la Recherche et de l'Industrie 1: 53-67, 1961. Paris: Centre de la Recherche Scientifique.

1191 _____. Direct Use of the Sun's Energy. New York: Ballantine Books, 1974. $1.95(pap). Copyright 1964 by Yale University, New Haven: Yale University Press, 1964. Tables, illus., diagrs., index. 271pp.

1192 _____. "Energy Efficiency in Photosynthesis." In Hollaender, A. Radiation Biology. New York: McGraw-Hill, 1956. pp. 259-92.

1193 _____. "Energy Storage Problems." United Nations Conference on New Sources of Energy, E35gr-Gen. 2, Rome, 1961. New York: United Nations, 1964. Also in Solar Energy 6: 78-83, 1962.

1194 _____. "Solar Energy." Science 109(2821): 51-57, Jan. 21, 1949.

1195 _____. "The Solar Era: Part II--Power Production with Small Solar Engines." Mechanical Engineering 94(9): 16ff (3pp.), Sept. 1972. Table, diagrs., refs.

1196 _____. "A Table of Quantum Yields in Experimental Photochemistry." Journal of Physical

Chemistry 42: 713-32, 1938.

1197 _____. Utilization of Solar Energy. Philadelphia: American Society for Testing Materials, 1960. 14pp. Illus.

1198 _____, and Alberty, R. A. Physical Chemistry (Chs. 18 and 23). New York: Wiley, 1961.

1199 _____, and Breihan, R. "Solar Focusing Collectors of Plastics." United Nations Conference on New Sources of Energy, E35-S104, Rome, 1961. New York: United Nations, 1964.

1200 _____, and Duffie, John A., eds. Solar Energy Research. Madison: University of Wisconsin Press, 1955. Illus., diagrs., tables. $10.00.

1201 _____, Williams, J. W., Bender, P., Alberty, R. A., and Cornwell, C. D. Experimental Physical Chemistry. New York: McGraw-Hill, 1962. 340pp.

1202 Daniels, Farrington, Jr. "Man and Radiant Energy: Solar Radiation." Handbook of Physiology. (Vol. 5: Environment). Washington, DC: American Physiological Society, 1963. pp. 969-87).

1203 Dannies, J. H. "Solar Air Conditioning & Solar Refrigeration." Solar Energy 3(1): 34-39, 1959.

1204 Davey, E. T. "Solar Water Heating in Australia." Paper presented at Melbourne International Solar Energy Society Conference, 1970.

1205 Davidson, H. L. "Horizontal Efficiency Coil Checker." Radio-Electronics 41: 67-68, May 1970. Illus., diagrs.

1206 Davies, J. M. and Cotton, E. S. "Design of the Quartermaster Solar Furnace." Solar Energy 1 (2-3): 16, 1957.

1207 Davis, C. P. and Lipper, R. I. "Solar Energy for Crop Drying." United Nations Conference on New Sources of Energy, E35-S53, Rome, 1961. New York: United Nations, 1964.

1208 "Dawn of an Industry. " Nations Business 62(9): 38-42, Sept. 1974. Photos.

1209 Deb, S. and Saha, H. "Secondary Ionization and Its Possible Bearing on the Performance of a Solar Cell. " Solid-State Electronics 15: 1389-91, Dec. 1972. Refs.

1210 DeBurchere and Alexandroff, G. "Introduction à l'Architecture Solaire. " L'Architecture d'Aujourd'-hui (140): 4-7, Oct. 1968. Illus., plans, diagrs.

1211 Degelman, L. O. The Development of a Mathematical Model for Predicting Solar Heat Gains through Building Walls and Roofs. University Park, PA: Pennsylvania State University, College of Engineering, Institute for Building Research, 1966. (Better Building Report #6). 55pp. Illus.

1212 Dehrer, L. G. "Solar Energy Heats Swimming Pool. " Heating, Piping, and Air Conditioning 41(10): 94-96, Oct. 1969. Photo., graph., diagrs.

1213 deJong, B. "Net Radiation Received by a Horizontal Surface at the Earth. " Monograph published by Delft University Press, Netherlands, 1973.

1214 DeKorne, James B. "Hot Water for the Homestead. " Mother Earth News (25): 72-79, Jan. 1974. Photos., plans.

1215 DelaRue, R. Lob, E. , Brenner, J. L. , and Hiester, N. K. "Flux Distribution Near the Focal Plane. " Solar Energy 1(2, 3): 94, 1957.

1216 Delyannis, A. A. and Delyannis, E. A. "Solar Desalting. " Chemical Engineering 77(22): 136-40, Oct. 19, 1970. Photos., table., diagrs.

1217 DeOnis, Juan. "Energy Crisis: Shortages and New Sources. " New York Times, p. 1, April 16, 1973; p. 1, April 17, 1973; p. 1, April 18, 1973. Diagr., graph, map, photos.

1218 Development of Gallium Arsenide Solar Cells (Final Report). Burlington, MA: Ion Physics Corp. , 1973. 43pp. NTIS #N73-30977/5GA. Also published,

Cincinnati, OH: Cincinnati University, 1973. 43pp.

1219 deWinter, F. "How to Design and Build a Solar-Energy Swimming Pool Heater." Publication of the Copper Development Association, Inc. 1973.

1220 Dictionary of World Activities and Bibliography of Significant Literature. Tempe, AZ: Association for Applied Solar Energy, 1959.

1221 Dietz, A. G. H. "Large Enclosures and Solar Energy." Architectural Design 41: 239-40, April 1971. Diagrs.

1222 Donovan, Paul and Woodward, William. An Assessment of Solar Energy As A National Energy Resource. College Park, MD: University of Maryland, Department of Mechanical Engineering, 1972. 89pp. NTIS #N73-26818/7.

1223 "Down on the Solar Farm." UNESCO Courier 27(1): 21, Jan. 1974.

1224 Drabkin, L. M. "Calculation and Cost Optimization of Certain Solar Generator Thermobattery Parameters." Geliotekhnika (1): 9-15, 1971. Trans. by Army Foreign Science and Technology Center, Charlottesville, VA.

1225 Dropkin, D. and Somerscales, E. "Heat Transfer by Natural Convection in Liquids Confined by Two Parallel Plates Which are Inclined at Various Angles with Respect to the Horizontal." ASME, Journal Heat Transfer 87: 77+, 1965.

1226 Drummond, A. J. "Sky Radiation, Its Importance in Solar Energy Problems." Transactions of the Conference on the Use of Solar Energy: The Scientific Basis 1: 113-31, 132-35, 1958.

1227 Dryden, H. L. and Doenkas, A. E. "Solar Energy in the Exploration of Space." Solar Energy (Special Issue) 5, Sept. 1961.

1228 Dubin, Fred S. "Energy Conservation Through Design." Professional Engineer 43(10): 21-25, Oct. 1973.

1229 Duerre, H. and Goergens, B. "Der Helios-Solarzel-
len-Generator." (The Helios Solar Cell Generator) Paper presented at the 5th Dglr. Annual
Meeting Berlin, Oct. 4-6, 1972. NTIS #N73-15084.

1230 Duffie, J. A. "New Materials in Solar Energy Utilization." United Nations Conference on New Sources
of Energy, E35-Gr-S12, Rome, 1961. New York:
United Nations, 1964.

1231 _____. "Reflective Solar Cooker Designs." Trans-
actions of the Conference on the Use of Solar
Energy: The Scientific Basis 3(2): 79. Tucson:
University of Arizona Press, 1958.

1232 _____, and Beckman, W. A. Solar Thermal Processes. New York: Wiley (Interscience), 1974.

1233 _____, Lappala, R. P., and Löf, G. O. G. "Plastics for Focussing Collectors." Solar Energy 1
(4): 79, 1957.

1234 _____, and Löf, G. O. G. "Focusing Solar Collectors for Power Generation." Paper presented at
World Power Conference, Melbourne, 1962.

1235 _____, Löf, G. O. G., and Beck, R. "Laboratory
and Field Studies of Plastic Reflector Solar Cook-
ers." United Nations Conference on New Sources
of Energy, E35-S87, Rome, 1961. New York:
United Nations, 1964.

1236 _____, and Sheridan, N. R. "Lithium Bromide-
Water Refrigerators for Solar Operation." Me-
chanical and Chemical Engineering Trans., Inst.
Engrs. Australia Mc-1: 79, 1965.

1237 _____, and _____. "The Part-Load Performance
of an Absorption Refrigerating Machine." Supplement au Bulletin de l'Institute International du Froid
380-87, 1965.

1238 _____, Smith, C., and Löf, G. O. G. "Analysis of
World Wide Distribution of Solar Radiation." Bul-
letin 21, Engineering Experiment Station. Madison:
University of Wisconsin, 1964.

1238a Duncan, R. T. "Solar Energy Will Cool and Heat Atlanta School. " ASHRAE Journal 16(9): 47-49. Sept. 1974.

1239 Dunkle, R. V. "Solar Water Distillation: The Roof Type Still and a Multiple Effect Diffusion Still. " International Developments in Heat Transfer, Conference at Denver, Part 5, 895, 1961.

1240 _____. "Thermal Radiation Tables and Applications. " Transactions ASME 76.

1241 _____, and Davey, E. T. "Flow Distribution in Absorber Banks. " Paper presented at Melbourne International Solar Energy Society Conference, 1970.

1242 Duwez, P. , et al. "Operation and Use of a Lens-Type Solar Furnace. " Transactions of the Conference on the Use of Solar Energy: The Scientific Basis 2(1B): 213-21. Tucson: University of Arizona Press, 1958.

1243 Dyson, Freeman J. "Energy in the Universe. " Scientific American 224(3): 50ff (10pp.), Sept. 1971. Charts, diagr. , photos.

1244 Eckert, J. A. , Kelley, B. P. , Willis, R. W. , and Berman, E. "Direct Conversion of Solar Energy on Earth, Now. " Proceedings of the Intersociety Energy Conversion Engineering Conference, (IECEC), pp. 372-75, 1973.

1245 Edlin, F. E. "Materiaux pour l'Utilization de l'Energie Solaire. " Applications Thermiques de l'Energie Solaire dans le Domaine de la Recherche et de l'Industrie 1: 539-63, 1961.

1246 _____. "Plastic Glazings for Solar Energy Absorption Collectors. " Solar Energy 2(2): 3-7, 1958.

1247 _____. "Selective Radiation Reflection Properties of the Fibrous Alkali Titanates. " Solar Energy Symposium, University of Florida, Gainesville, Florida, April 1964. Transaction of American Society of Mechanical Engineers.

1248 _____. "Worldwide Progress in Solar Energy. "

Proceedings of the Intersociety Energy Conversion Energy Conference (IECEC), pp. 92-97, New York, 1968.

1249 _____, and Willauer, D. E. "Plastic Films for Solar Energy Applications." United Nations Conference on New Sources of Energy, E35-S33, Rome, 1961. New York: United Nations, 1964.

1250 Edmondson, W. B., ed. "Breakthrough in Selective Coatings." Solar Energy Digest 2(2): 1-2, Feb. 1974.

1251 Edwards, D. K. "Suppression of Cellular Convection by Lateral Walls." Transactions ASME, Journal Heat Transfer 91: 145+, 1969.

1252 _____, Gier, J. T., Nelson, K. R., and Roddick, R. D. "Spectral and Directional Thermal Radiation Characteristics of Selective Surfaces for Solar Collectors." Solar Energy 6(1): 1-8, 1962.

1253 _____, and Nelson, K. E. "Radiation Characteristics in the Optimization of Solar Heat-Power Conversion Systems." Paper 61-WA-158, presented at the 1961 Winter ASME Meeting.

1254 _____, et al. "Basic Studies on the Use and Control of Solar Energy." Annual Report NSF G9505 August 1959-August 1960. Los Angeles: Department of Engineering, University of California, 1960. Illus., refs.

1255 Egli, P. H. Thermoelectricity. New York: Wiley, 1960.

1256 Ehricke, K. A. "Extraterrestrial Imperative." Bulletin of the Atomic Scientists 27(9): 18ff (6pp.), Nov. 1971.

1257 Eibling, James A. "A Survey of Solar Collectors." Proceedings of the NSF/RANN Solar Heating and Cooling for Buildings Workshop, NSF/RANN-73-004, pp. 47-51, July 1973.

1258 _____, Thomas, R. E., and Landry, B. A. "An Investigation of Multiple-Effect Evaporation of

Saline Waters by Steam from Solar Radiation. "
Report from Battelle Memorial Institute to U. S.
Department of Interior, Dec. 1953.

1259 _____, _____, and _____. "An Investigation
of Multiple-Effect Evaporation of Saline Waters by
Steam from Solar Radiation. " Report to the Office
of Saline Water, U. S. Dept. of Interior, 1953.

1260 _____, et al. "Solar Stills for Community Use--
Digest of Technology. " Solar Energy 13(2): 263ff
(13pp.), May 1971. Charts, diagrs.

1261 Eisenstadt, M. M. , Flanigan, F. M. , and Farber,
E. A. "Solar Air Conditioning with an Ammonia-
Water Absorption Refrigeration System. " American
Society of Mechanical Engineers, Paper 59-A-276,
1959.

1262 _____, _____, and _____. "Tests Prove
Feasibility of Solar Air Conditioning. " Heating,
Piping and Air Conditioning 32(11): 120, 1960.

1263 "Electric Power From Space. " Mechanical Engineering
94(10): 51, Oct. 1972. Drawing.

1264 "Electricity Generation in Australia. " Search 2(9):
238ff (8pp.), Sept. 1971.

1265 Ellis, B. and Moss, T. S. "Calculated Efficiencies of
Practical Ga As and Si Solar Cells Including the
Effect of Built-In Electric Fields. " Solid State
Electronics 13: 1-24, June 1970. Refs. , diagrs.

1266 Elsner, M. K. and Kahlil, A. M. "Solar Radiation
Availability in the United Arab Republic. " United
Nations Conference on New Sources of Energy, E35-
S62, Rome, 1961. New York: United Nations, 1964.

1267 Elson, B. M. "Solar Array Capabilities Increase. "
Aviation Week & Space Technology 95(11): 50-51+,
Sept. 13, 1971. graph, photos.

1268 _____. "Theoretical Picture of Sun Still Evolving. "
Aviation Week and Space Technology 63: Jan. 14,
1974.

1269 "Emphasis Grows on Solar Energy. " Industrial Research 15(9): 23-24, Sept. 1973.

1270 "Energia Solare Come Alternativa? Progetto di Città Solare. " Casabella No. 374: 50-54, 1973. Illus., diagrs. (in Italian).

1271 "Energy Alternative for Aerospace. " Industrial Research 16: 29-30, March 1974.

1272 "Energy and Environment. " Bulletin of the Atomic Scientists 18(5): 5-8, May 1972.

1273 "Energy Conversion and Environmental Pollution. " Science Teacher 39(3): 31ff (5pp.), March 1972.

1274 "Energy Crisis: Are We Running Out?" Time 99(24): 49-53, June 12, 1972. Graph, photos.

1275 "Energy Crisis Perils Economic Growth. " Chemical and Engineering News 50(1): 4, Jan. 3, 1972. Photo.

1276 "Energy Developments: Solar Energy. " Congressional Quarterly 31(27): 1817, July 7, 1973.

1277 "Energy from Above and Below. " Nature 242(5392): 6-8, March 2, 1973.

1278 "Energy in the Year 2000. " Power from Oil p. 3-6, Fall 1973. Photos.

1279 "Energy in the Years 1975-2000. " Battelle Research Outlook 4(1): 2-6, 1972.

1280 "Energy in Transition to Steady State. " Presented at AAAS 138th Meeting, Dec. 26-31, 1971. Philadelphia. 17pp.

1281 "Energy Options for the United States. " IEEE Spectrum 10(9): 63ff (6pp.), Sept. 1973. Graphs, table.

1282 Energy Primer: Solar, Water, Wind, & Bio Fuels. Berkeley, CA: Portola Institute, 1974. $4. 50pap.

1283 "Energy Research and Development. " Report Hs Comm Science Astronautics 92 Con II Serial EE, Dec. 1972. 418pp.

1284 "Energy Research Needs: Residential Utilization Technology, VIII." NTIS Report PB-207 516, Oct. 1971. 52pp.

1285 Engebretson, C. D. "Use of Solar Space Heating, M. I. T. House IV." United Nations Conference on New Sources of Energy, E35-S67, Rome, 1961. New York: United Nations, 1964.

1286 _____, and Ashar, N. G. "Progress in Space Heating with Solar Energy." Paper 60-WA-88 presented at the New York ASME Meeting, 1960.

1287 "Engineering Experimental Program On the Effects of Near Space Radiation on Lithium-Doped Solar Cells." Semiannual Report July 2, 1969-April 1970. Palo Alto, CA: Philco-Ford, 1970. 54pp. NTIS #N-70-42732/GA. (Final Report), 1971, 109pp. #N-72-22043/GA.

1288 Engineering of Thermoelectricity. New York: Interscience, 1961.

1289 Engineering Societies Library. Bibliography on Domestic and Industrial Applications of Solar Heating. New York: Engineering Societies Library, 1950.

1290 Epstein, A. S., et al. "Effect of Oxygen on GaP Solar Cell Characteristics." Solid State Electronics 14: 757-60, Aug. 1971.

1291 Erricson, J. "Sun Power: The Solar Engine." Contributions to The Centennial, pp. 571-77, Philadelphia, 1870.

1292 Evans, D. G. and McCutchan, J. C. "Electricity Generation in Australia." Search 2(9): 338ff (8pp.), Sept. 1971. Maps, charts, graphs.

1293 Evans, G. E. "Electrochemical Energy Storage for Intermittent Power Sources." United Nations Conference on New Sources of Energy, E35-Gen. 3, Rome, 1961. New York: United Nations, 1964.

1293a Ewer, William. How to Use Solar Energy. Phoenix, AZ: Sincere Press, [1975?]. $8.95 (also in pap).

1294 "Experimental Solar Home Makes Fair Debut, Begins Two-Year Test." Machine Design 46(23): 6, Sept. 19, 1974.

1294a Fairbanks, J. W. and Morse, F. H. "Passive Solar Array Orientation Devices for Terrestrial Application." Solar Energy 14: 67-69, 1972.

1295 Farber, Erich A. "A Closed Cycle Solar Engine." Solar Energy Symposium, University of Florida, Gainesville, 1964.

1296 _____. "Design and Performance of a Compact Solar Refrigeration System." Engineering Progress at the University of Florida 24(2): 70, 1970.

1297 _____. "An Open Cycle Solar Engine." Solar Energy Symposium, University of Florida, Gainesville, Florida, 1964.

1298 _____. "Solar Energy: Conversion and Utilization." Building Systems Design 69(6): 25ff (9pp.), June 1972. Photos. Also in The Nucleus 5(1-2): 14-23, Jan.-June 1963.

1299 _____. "Sun Power Harnessed to Run Equipment at Solar Energy Lab." Mechanical Engineering 92(4): 82-83, April 1970. Photo. diagrs.

1300 _____. "The Use of Solar Energy for Heating Water." United Nations Conference on New Sources of Energy, E35-S1, Rome, 1961. New York: United Nations, 1964.

1301 _____, Flanigan, F. M., Lopez, L., and Politaka, R. W. "University of Florida Solar Air-Conditioning System." Solar Energy 10(2): 91-95, April 1966.

1302 _____, and Prescott, F. L. "Closed Cycle Solar Hot Air Engines." Solar Energy 9(4): 170-74, Oct. 1965.

1303 _____, and _____. Solar Engines. Gainesville, FL: Florida University, Gainesville Engineering and Industrial Experiment Station, Technical Progress Report #14. 1965. 12pp. Refs., illus.

1304 Farber, J. "Utilization of Solar Energy for the Attainment of High Temperatures." Solar Energy Research (Daniels, F. and Duffie, J. A., eds.) Madison: University of Wisconsin Press, pp. 157-61.

1305 Farnham, Lee L., Stewart, John K., and Petrou, Nicholas. "Solar Heating and Cooling of Buildings --Problems of Commercialization." Energy Environment Productivity, Proceedings of the First Symposium on Rann: Research Applied to National Needs, Washington, D.C., Nov. 18-20, 1973. Washington: National Science Foundation, pp. 54-59.

1306 Finklestein, T. "Cyclic Processes in Closed Regenerative Gas Machines." American Society of Mechanical Engineers, Paper 61-SA-21, Annual Meeting, 1961.

1307 _____. "Generalized Thermodynamic Analysis of Stirling Engines." American Society of Automotive Engineers, Paper 118 B., Annual Meeting, 1960.

1308 _____. "Internally Focusing Solar Power Systems." American Society of Mechanical Engineers, Paper 61-WA-297, Annual Meeting, 1961.

1309 Fischer, Horst and Gereth, Reinhard. "Telesun the Solar Cell of the Future." AEG Telefunken Progress 1: 25-26, 1972.

1310 Fisher, A. W., Jr. "Economic Aspects of Algae as Potential Fuel." Solar Energy Research (Daniels, F. and Duffie, J. A. eds.) Madison: University of Wisconsin Press, 1955. pp. 185-89.

1311 _____. "Engineering for Algae Culture." Proceedings, World Symposium on Applied Solar Energy, Phoenix, AZ, 1955, pp. 243-54. Menlo Park, CA: Stanford Research Institute, 1956.

1312 Fisher, Fred D. Construction and Operation of a High Temperature Solar Furnace and Studies of Elemental Boron. Oregon State #60-3336.

1313 Fisher, J. H. "An Analysis of Solar Energy Utilization." WADC Technical Report 59-17, Vols. 1 and

2, Wright Air Development Center, ASTIA Document No. AD 214611, 1959.

1314 Fitzmaurice, R. and Seligman, A. C. "Some Experiments on Solar Distillation of Sea Water in Cyprus." Transactions of the Conference on the Use of Solar Energy: The Scientific Basis 3: 109-18. Tucson: University of Arizona Press, 1958.

1315 "Flexible Solar-Cell Arrays." Electronics 43(11): 70-71, May 25, 1970.

1316 "Floating Station Collects Sunlight." Industrial Research 14: 32, June 1972.

1317 Foex, M. "Mesure des Temperatures au Four Solaire." United Nations Conference on New Sources of Energy E35-S, Rome, 1961. New York: United Nations, 1964.

1318 Fontan, L. and Barasoain, J. A. "Some Examples of Small-Scale Solar Distillation of Water." United Nations Conference on New Sources of Energy, E35-S73, Rome, 1961. New York: United Nations, 1964.

1319 Ford, Norman C. and Kane, Joseph W. "Solar Power." Bulletin of Atomic Scientists 27(8): 27ff (5pp.), Oct. 1971. Drawings, tables, portrait.

1320 "Forum: Energy's Impact on Architecture." Architectural Forum 139(1): 59ff (16pp.), July-Aug. 1973. Photos.

1321 Frank, Walther, et al. Sonneneinstrahlung: Fenster, Raumklime, Untersuchungen Durchgefuhrt im Auftrage des Bundes-Ministers Fur Stadtban und Wohnungswesen und der Huttentechnischen Vereinigung der Deutschen Glasindustrie, e.V., Frankfurt (Main), Berkin: W. Ernst, 1970. Illus.

1322 Free, John R. "Fast New Process Forms Sapphires and Solar Cells." Popular Science 205(3): 69-71, Sept. 1974. Photos, diagrs.

1322a _____. "Solar Cells: When Will You Plug Into Electricity from Sunshine?" Popular Science 205

(6): 53ff (4-1/2pp.), Dec. 1974. Photos., plans, diagrs.

1323 Freeman, S. David. "The Energy Joyride is Over."
      Bulletin of Atomic Scientists 29(8): 39-41, Oct. 1973.

1324 "French Switch on to Sun Power." Business Week
      (2123): 126-29, May 9, 1970. Photos.

1325 Friedlander, Gordon D. "Energy: Crisis and Chal-
      lenge." IEEE Spectrum 10(5): 18ff (10pp.), May
      1973. Diagr., graphs, photos., table.

1326 Frisch, Joan Vandiver. "More Sunshine in Your Life."
      NOAA (492): 10ff (4pp.), April 1974. Photos.

1327 Fritz, S. "Some Solar Radiation Data Presentations
      for Use in Applied Solar Energy Programs." (by
      an international group of authors including S. Fritz,
      E. Barry, H. M. P. Manolova, and T. H. Mac-
      Donald). Solar Energy 4(1): 1-22, 1960.

1328 _____. "Solar Radiation Energy and Its Modification
      by the Earth and Its Atmosphere." Compendium of
      Meteorology, American Meteorological Society, 1951.

1329 _____. "Transmission of Solar Energy Through the
      Earth's Clear and Cloudy Atmosphere." Transac-
      tions of the Conference on the Use of Solar Energy:
      The Scientific Basis 1: 17-36, 1958.

1330 Frost, George R. From Sun to Sound. New York:
      Bell Telephone Laboratories, Inc., 1961. 117pp.
      Illus., portraits, diagrs.

1331 Fry, J. and Nicoletta, C. A. Ultraviolet and Charged
      Particle Irradiation of Proposed Solar Cell Cover-
      slide Materials and Conductive Coatings for the
      Helios Spacecraft. Greenbelt, MD: Goddard Space
      Flight Center, 1972. 112pp. NTIS #N73-18571.

1332 Fuel Cell Technical Manual. New York: Publication
      Department American Institute of Chemical Engineer,
      1963.

1333 Fukuo, N., et al. "Installations for Solar Space Heat-
      ing in Gerin." United Nations Conference on New

Sources of Energy, E35-S112, Rome, 1961. New York: United Nations, 1964.

1334 Gallager, Sheldon M. "Switching on the Sun: First Solar Home to Produce Both Heat and Electricity." Popular Mechanics 140(4): 190-92, Oct. 1973. Photos., diagrs.

1335 Gannon, Robert. "Atomic Power, What Are Our Alternatives?" Science Digest 70(6): 18ff (5pp.), Dec. 1971. Photos.

1336 Gardner, A. L. "Articulated and Semi-Articulated Low-Cost Hot Focus Solar Energy Concentrators." New Delhi Symposium on Solar Energy and Wind Power, pp. 209-15, Paris: UNESCO, 1956.

1337 Garg, H. P. "Design and Performance of a Large-Size Solar Water Heater." Solar Energy 14: 303-12, 1973.

1338 _____, and Gupta, C. L. "Design Data for Direct Solar Utilization Devices--1, System Data." Journal of the Institute of Engineers (India), (Special Issue) 48: 461, 1968.

1339 _____, and _____. "Design of Flat-Plate Solar Collectors for India." Journal of the Institute of Engineers (India), (Special Issue) 47: 382, 1967.

1340 Gates, David M. "The Energy Environment in Which We Live." American Scientist 5: 327-48, 1963.

1341 _____. "The Flow of Energy in the Biosphere." Scientific American 224(3): 88ff (11pp.), Sept. 1971. Map, charts, photo.

1342 Gaucher, Leon P. "Energy in Perspective." Chemical Technology, March 1971.

1343 _____. "Energy Requirements of the Future." Solar Energy 14(1): 5ff (5pp.), Dec. 1972. Diagr., graph, refs.

1344 _____. "Energy Sources of the Future for the United States." Solar Energy 9(3): 119-26, 1965.

1345 _____. "The Solar Era: Part I--The Practical
       Promise. " Mechanical Engineering 94(8): 9-12,
       Aug. 1972. Graphs, charts, refs.

1346 General Electric Research Bulletin, Summer 1963. p. 4.

1347 Geoffrey, J. "Use of Solar Energy for Water Heat-
       ing. " United Nations Conference on New Sources
       of Energy, E35-S58, Rome, 1961. New York:
       United Nations, 1964.

1348 "Getting More from the Sun. " Industrial Research 14:
       27, July 1972.

1349 Ghai, M. L. "Applications of Solar Energy for Heat-
       ing, Air Conditioning, and Cooling. " Solar Energy
       Research (Daniels, F. and Duffie, J. A. , eds.).
       Madison: University of Wisconsin Press, 1955.

1350 "Giant Solar Panel Will Power Space Stations. " Pro-
       duct Engineering 41(18): 10, Aug. 31, 1970.

1351 Gier, J. T. and Dunkle, R. V. "Selective Spectral
       Characteristics as an Important Factor in the Effi-
       ciency of Solar Collectors. " Transactions of the
       Conference on the Use of Solar Energy: The Scien-
       tific Basis 2(1A): 41-56. Tucson: University of
       Arizona Press, 1958.

1352 Gill, W. D. and Bube, R. H. "Light Microprobe In-
       vestigation of CU_2S-CdS Heterojunctions. " Journal
       of Applied Physics 41(4): 1694-1700, March 15,
       1970. Refs. , graphs. , illus. , diagrs.

1353 Gillette, R. B. "Analysis of the Performance of a
       Solar Heated House. " M. S. Thesis in Mechanical
       Engineering. Madison: University of Wisconsin,
       1959.

1354 _____. "Selectively Emissive Materials for Solar
       Heat Absorbers. " Solar Energy 4(4): 24-32, 1960.

1355 _____, Snyder, H. E. , and Timar, T. "Light-
       weight Solar Concentrator Development. " Solar
       Energy 5(1): 24-28, 1961.

1356 Gilman, Stanley F. and Sturz, Douglas, H. "Solar

Energy Assisted Heat Pumps Systems for Commercial Office Buildings. " Papers presented at
the Symposium on Solar Energy Applications at the
ASHRAE Annual Meeting, June 23-27, 1974, Montreal, Canada. pp. 26-34. Refs., tables, graphs.

1357 Gilmore, C. P. "Can Sunshine Heat (And Cool) Your
House?" Popular Science 204(3): 78-81+, March
1974. Photos, diagrs.

1358 _____. "On the Way: Plentiful Energy from the
Sun. " Popular Science 201(6): 86-9+, Dec. 1972.
Photos, diagrs., drawings.

1359 Girardier, J. P., Alexandroff, G., Alexandroff, J.
"Les Moteurs Solaires et l'Habitat pour les Zones
Arides. " Paper presented at International Solar
Energy Congress, Paris, 1973.

1360 Giutronich, J. E. "The Design of Solar Concentrators
Using Toroidal, Spherical, or Flat Components. "
Solar Energy 7(4): 162-66, 1963.

1361 Glaser, Peter E. "A New View of Solar Energy. "
Proceedings of the Intersociety Energy Conversion
Engineering Conference (IECEC), paper no. 719002,
pp. 1-4, 1971.

1362 _____. "Beyond Nuclear Power--The Large Scale
Use of Solar Energy. " Transactions of the New
York Academy of Sciences, 1969.

1363 _____. "Concept for a Satellite Solar Power System. " Chemical Technology 1: 606-14, Oct. 1971.

1364 _____. "Engineering Research with a Solar Furnace. " Solar Energy 2(2): 7-10, 1958.

1365 _____. "Industrial Applications--The Challenge to
Solar Furnace Research, #25. " United Nations
Conference on New Sources of Energy, E35-S,
Rome, 1961. New York: United Nations, 1964.

1366 _____. "New Look at the Sun. " Conservationist
25(6): 22-27, June 1971. Drawing, photos.

1367 _____. "Power from the Sun. " UNESCO Courier

27(1), 16-21, Jan. 1974. Photos., drawings.

1368 _____. "Power from the Sun: Its Future." Science 162: 857-61, Nov. 22, 1968.

1369 _____. "Satellite Solar Power Station." Solar Energy 12(3): 353-61, 1969.

1370 _____. "Solar Energy--Future Power Source." Petroleum Engineer International 44: EM 6-9, Jan. 1973. Also in Pipeline and Gas Journal 199: EM 6-9, Jan. 1972.

1371 _____. "Solar Energy--Prospects for Its Large-Scale Use." Science Teacher 39(3): 36-40, March 1972. Drawings.

1372 _____. "Solar Power via Satellite." Astronautics & Aeronautics 11(8): 60ff (9pp.), Aug. 1973. Diagrs., drawing, graphs, photos., refs., tables.

1373 _____. "The Future of Power from the Sun." Power 112(8): 61-63. Also from Proceedings Intersociety Energy Conversion Engineering Conference (IECEC), pp. 98-103, 1968.

1374 _____, and Burke, James C. "New Directions for Solar Energy." Bulletin of the Atomic Scientists 29(8): 40-43, Oct. 1973.

1375 _____, Maynard, O. E., Mockavciak, J., and Ralph, E. L. "Feasibility Study of a Satellite Solar Power Station." NASA Contractor Report CR-2357, Feb. 1974.

1376 _____, and Walker, Raymond F. Thermal Imaging Techniques. New York: Plenum Press, 1964. $20.00.

1377 "Glass Tubing, Liquid Sodium Extract Energy from the Sun." Engineering News-Record 186(22): 14, June 3, 1971.

1378 Glitch, Rolf E. "UNESCO and the Changing Attitude Toward Solar Energy." UNESCO Courier 27(1): 14-15, Jan. 1974. Photo.

1379 Golueke, C. G. and Oswald, W. J. "Power from
       Solar Energy via Algae-Produced Methane. " Solar
       Energy 7(3): 86-92, July, 1963.

1380 _____, and _____. "Recycling System for Single-
       Family Farms and Villages. " Organic Farming and
       Gardening 20(8): 64-66, Aug. 1973. Diagr.

1381 Goldberg, Michael A. "The Economics of Limiting
       Energy Use. " Alternatives 1(4): 3ff (11pp.), Sum-
       mer 1972. Graphs, photo.

1382 Goldsmid, H. J. Applications of Thermoelectricity.
       London: Methuen, 1960.

1383 Goldstein, M. "Some Physical-Chemical Aspects of
       Heat Storage. " United Nations Conference on New
       Sources of Energy, E35-S7, Rome, 1961. New
       York: United Nations, 1964.

1384 Gomella, C. "Possibilities of Increasing the Dimen-
       sions of Solar Stills. " United Nations Conference
       on New Sources of Energy, E35-S107, Rome, 1961.
       New York: United Nations, 1964.

1385 _____. "Practical Possibilities for the Use of Solar
       Distillation in Underdeveloped Arid Countries. "
       Transactions of the Conference on the Use of Solar
       Energy: The Scientific Basis 3: 119-33, Tucson:
       University of Arizona Press, 1958.

1386 _____. "Use of Solar Energy for the Production of
       Fresh Water. " United Nations Conference on New
       Sources of Energy, E35-Gr-s19, Rome, 1961. New
       York: United Nations, 1964.

1387 Gould, William R. "Energy Options for the Future. "
       Speech presented at the National Science Teachers
       Association Convention, Detroit, April 2, 1973.
       13pp. Vital Speeches of the Day 39(19): 605-08,
       July 15, 1973.

1388 Gräfe, K. "Measurements of Total Radiation in Net-
       work Networks. " United Nations Conference on New
       Sources of Energy, E35-61, Rome, 1961. New
       York: United Nations, 1964.

1389 Graham, J. D. "Solar Induced Binding Vibrations of a
      Flexible Member." AIAA Journal 8: 2031-36,
      Nov. 1970. Refs., diagrs.

1390 Graven, R. M. Calculations On a Solar Energy Sys-
      tem. Berkeley, CA: University of California,
      Lawrence Lab, 1973. 33pp. NTIS #LBL-1773/GA.

1391 Green, John M. Lithium-doped Silicon Solar Cells
      State-of-the-Art, Wright-Patterson AFN, OH: Air
      Space Force Aero Propulsion Lab, 1972. 42pp.
      NTIS #AD-764357.

1392 Green, W. P. Utilization of Solar Energy for Air
      Conditioning and Refrigeration in Florida. Masters
      Thesis, College of Engineering, University of
      Florida, Gainesville, 1936.

1392a Greene, A. M. "Solar Concentrators: New Energy
      Alternative." Iron Age 214(16): 65-67, Oct. 14,
      1974. Photos., diagrs.

1393 Greenwood, R. F. Results of the 1970 Balloon Flight
      Solar Cell Standardization Program. Pasadena,
      CA: Jet Propulsion Lab, California Institute of
      Technology, 1972. 21pp. NTIS #N73-13050.

1394 Grey, Jerry. "Prospecting for Energy." Astro-
      nautics & Aeronautics 11(8): 24-29, Aug. 1973.
      Drawings, refs.

1395 Grier, N. T. and McKinzie, D. J., Jr. "Current
      Drainage to a High Voltage Probe in a Dilute
      Plasma." Journal of Spacecraft and Rockets 9:
      706-08, Sept. 1972. Refs.

1396 Grodzka, P. G. "Thermal Control of Spacecraft by
      Use of Solid-Liquid Phase-Change Materials."
      AIAA 8th Aerospace Sci Meeting, NY: Jan. 1970.

1397 Grune, W. N., Collins, R. A., and Thompson, T. L.
      "Forced Convection, Multiple-Effect Solar Still for
      Desalting Sea and Brackish Waters." United Nations
      Conference on New Sources of Energy, pp. 1-26,
      E35-S14, Rome, 1961. New York: United Nations,
      1964.

1398 Guner, K. A., et al. "The Application of Phenol-
      Formaldehyde Plastic Foam as Thermal Insulation
      for Solar Water Heaters." Geliotekhnika (4), 1970.
      Trans. by Army Foreign Science & Technology
      Center, Charlottesville, VA. NTIS #AD-756 591.

1399 Gunter, Carl. "The Utilization of Solar Heat for In-
      dustrial Purposes by Means of a New Plane Mirror
      Reflector." Scientific American, May 26, 1906.

1400 Gupta, G. L. "On Generalizing the Dynamic Perform-
      ance of Solar Energy Systems." Solar Energy 13:
      301, 1971.

1401 _____, and Garg, H. P. "System Design in
      Solar Water Heaters with Natural Circulation."
      Solar Energy 12: 163-82, 1968.

1402 Gutierrez, G., Hincapie, F., Duffie, J. A., and Beck-
      man, W. A. "Simulation of Forced Circulation
      Water Heaters; Effects of Auxiliary Energy Supply,
      Load Type, and Storage Capacity." Solar Energy
      15: 287, 1974.

1403 Hafez, M. M. and Elnser, M. K. "Demineralization
      of Saline Water by Solar Radiation in the United
      Arab Republic." United Nations Conference on New
      Sources of Energy, E35-S63, Rome, 1961. New
      York: United Nations, 1964.

1404 Halacy, Daniel Stephen, Jr. The Coming Age of Solar
      Energy. New York: Harper and Row, 1973.
      231pp. Illus. (junior high school & up).

1404a _____. Experiments with Solar Energy. New York:
       W. W. Norton & Co., 1969. Illus., refs., index.
       $4.50. PLB $4.14. (Gr. 5-9). 147pp.

1405 _____. Fabulous Fireball: The Story of Solar
      Energy. New York: Macmillan Co., 1957. 154pp.
      Illus. (gr. 7 and up). $4.95.

1406 _____. Fun with the Sun: 7 Solar Energy Projects
      to Build and Use. New York: Macmillan, 1961.
      Illus., photos., diagrs. 112pp. (junior high
      school and up).

1407 _____. "How to Build--and Use!--a Reflector
Cooker." Mother Earth News (26): 68-71, March
1974. Drawings, plans, diagrs.

1408 _____. "How to Build--and Use!--a Solar Oven."
Mother Earth News (25): 26-28, Jan. 1974.
Photos., plans.

1409 _____. "The Solar Alternative to Atomic Power."
Science Digest 71(3): 45ff (6pp.), March 1972.
Photos.

1410 _____. "A Solar Furnace: How to Build and Use
One." The Mother Earth News (28): 72-74, July
1974.

1411 _____. Solar Science Projects. New York: Scho-
lastic Book Service, 1971. 96pp.

1412 Halb, D. F., et al. "Electrostatic Rocket Exhaust
Effects on Solar-Electric Spacecraft Subsystems."
Journal of Spacecraft and Rockets 7: 305-12, March
1970. Refs., diagrs.

1413 Hale, D. V., Hoover, M. J., and O'Neill, M. J.
Phase-Change Materials Handbook NASA CR-61363,
Sept. 1971.

1414 Hamilton, [U. S. Rep.] Lee H. "The Energy Crisis:
The Persian Gulf." Vital Speeches 39(8): 226-31,
Feb. 1, 1973.

1415 Hamilton, Richard W., ed. Space Heating with Solar
Energy: Proceedings of a Course-Symposium Held
at the Massachusetts Institute of Technology, Aug.
21-26, 1950. Cambridge: Massachusetts Institute
of Technology, 1954. 161pp. Illus., maps, diagrs.
54-4959.

1416 Hammond, Allen L. "Energy Options: Challenge for
the Future." Science 177(4052): 875-877, Sept. 8,
1972. Charts.

1417 _____. "Individual Self-Sufficiency in Energy."
Science 184(4134): 278-82, April 19, 1974. Diagrs.,
photos., refs.

1418 _____. "Photovoltaic Cells: Direct Conversion of Solar Energy." Science 178(4062): 732-34, Nov. 17, 1972.

1419 _____. "Solar Energy: A Feasible Source of Power?" Science 172(3984): 660, May 14, 1971.

1420 _____. "Solar Energy: Proposal for a Major Research Program." Science 179(4078): 1116, March 16, 1973.

1421 _____. "Solar Energy: The Largest Resource." Science 177 (4054): 1088-90, Sept. 22, 1972.

1422 _____. "Solar Power: Promising New Developments." Science 184(4144): 1359-60. June 28, 1974. Photo., diagr.

1423 Hand, I. F. "Charts to Obtain Solar Altitudes and Azimuths." Heating and Ventilating 45: 86, Oct. 1948.

1424 _____. "Insolation on Clear Days at the Time of Solstices and Equinoxes for Latitude 42°N." Heating and Ventilating 47: 92, 1950.

1425 _____. Monthly Weather Review 65: 415-41, 1937: Also in U.S. Weather Report, Nov. 1946.

1426 Handley, D. and Heggs, P. S. "The Effect of Thermal Conductivity of the Packing Material on the Transient Heat Transfer in a Fixed Bed." International Journal of Heat Mass Transfer 12: 549, 1969.

1426a "Harnessing the Sun." Forbes 114(8): 23-24, Oct. 15, 1974. Photo.

1427 Harris, L., McGinnes, R. T., and Siegel, B. M. "The Preparation and Optical Properties of Gold Blacks." Journal of the Optic Society of America 38: 582-89, 1948.

1427a Harwood, Michael. "Energy From Our Star Will Compete with Oil, Natural Gas, Coal, and Uranium, But Not Soon." New York Times Magazine, March 16, 1975, p. 42-49. Photos., diagrs.

1428 Hasbach, W. A. and Ross, R. G., Jr. The Develop-
       ment, Design, and Test of a 6W/K_G (30-W/lb)
       Roll-Up Solar Array. Pasadena, CA: Jet Propul-
       sion Lab California Institute of Technology, 1972.
       38pp. NTIS #N72-32070.

1429 Hatsopoulos, G. N. and Brosens, P. J. "Solar-
       Heated Thermoionic Converter." United Nations
       Conference on New Sources of Energy, E35-S78,
       Rome, 1961. New York: United Nations, 1964.

1430 Hauser, L. G. "Assessing Advanced Methods of Gen-
       eration." Electrical Review 193(7): 219-23, Aug.
       17, 1973. Diagr.

1431 Hay, Harold R. "Energy, Technology and Solarchi-
       tecture." Mechanical Engineering 95(11): 18ff
       (5pp.), Nov. 1973. Photos.

1432 _____. "Evaluation of Proved Natural Radiation
       Flux Heating and Cooling." Proceedings of the
       NSF/RANN Solar Heating and Cooling for Build-
       ings Workshop, NSF/RANN-73-oo4, pp. 185-87,
       July 1973.

1433 _____. "New Roofs for Hot Dry Regions." Ekis-
       tics 31: 158-64, 1971.

1434 _____. "The Solar Era, Part 3--Solar Radiation:
       Some Implications and Adaptations." Mechanical
       Engineering 94 (10): 24ff (6pp.), Oct. 1972. Diagrs.,
       drawings, graph, photo., refs.

1435 _____, and Yellot, J. I. "A Naturally Air-Condi-
       tioned Building." Mechanical Engineering 92(1):
       19-25, Jan. 1970.

1436 _____, and _____. "Natural Air Conditioning
       with Roof Ponds and Movable Insulation." ASHRAE
       Transactions 75(1): 165-77, 1969.

1437 "Heating Up and Cooling Off Down Under." Progres-
       sive Architecture 52(10): 32, Oct. 1971.

1438 Hedger, John H. The Sleeping Giant: Solar Energy.
       Lakeside, CA: Hedger Scientific Specialties, 1963.
       126pp. Illus., facsims., portraits.

1439 Heidt, L. J., Livingston, R. S., Rabinowitch, E., and Daniels, F. Photochemistry in the Liquid and Solid States. New York: Wiley, 1960.

1440 Heiserman, D. L. "Power from the Sun with Silicon." Popular Electronics (including Electronics World) 3(2): 69-71, Feb. 1973. Photo, diagr.

1441 Henahan, John F. "How Soon the Sun: A Timetable for Solar Energy." Saturday Review/World 1(6): 64-66, Nov. 20, 1973.

1442 _____. "Timetable for Solar Energy." Current (158): 3-7, Jan. 1974.

1443 Henry, O. "House that has its Furnace in the Sky." Popular Mechanics 139(6): 150-54, June 1973.

1444 "Here Is a Brief Rundown on Ways Electricity is Generated." Aware (27): 3-10, Dec. 1972.

1445 Heronemus, William, E. "Alternatives to Nuclear Energy." Catalyst for Environmental Quality 2(3): 21-25, 1972. Drawing, photo.

1446 Heywood, H. "The Computation of Solar Radiation Intensities." Solar Energy 10: 51+, 1966.

1447 _____. "Mechanical Construction and Thermal Characteristics of Solar-Operated Thermoelectric Generators." Transactions of the Conference on the Use of Solar Energy: The Scientific Basis 5(1-7): 8-20. Tucson: University of Arizona Press.

1448 _____. "Simple Instruments for the Assessment of Daily Solar Radiation Intensity." United Nations Conference on New Sources of Energy, E35-S9, Rome, 1961. New York: United Nations, 1964.

1449 Hibbard, R. R. "Equilibrium Temperatures of Ideal Spectrally Selective Surfaces." Solar Energy 5(4): 129-32, 1961.

1449a "High Oil Costs Heat Up U.S. Solar Energy Research." Engineering News-Record 193(21): 32-42, Nov. 15, 1974.

1450 Hildebrandt, A. F., et al. "Large-Scale Concentra-
      tion and Conversion of Solar Energy." American
      Geophysical Union, Transactions 53(7): 684ff (9pp.),

1451 _____, and Vant-Hull, L. L. "Large-Scale Utiliza-
      tion of Solar Energy." Energy Research and De-
      velopment; Hearings of the House Committee on
      Science and Astronautics, pp. 499-505. Washing-
      ton: U. S. Government Printing Office, 1972.

1452 Hill, R. "Oxygen Evolved by Isolated Chloroplasts."
      Nature 139: 881-82, 1937.

1453 Hirschmann, J. G. "A Solar Energy Pilot Plant for
      Northern Chile." Solar Energy 5: 37-43, 1961.

1454 _____. "Solar Evaporation and Distilling Plant in
      Chile." United Nations Conference on New Sources
      of Energy, E35-S23, Rome, 1961. New York:
      United Nations, 1964.

1455 Hisada, T. "Construction of a Solar Furnace."
      Transactions of the Conference on the Use of Solar
      Energy: The Scientific Basis 2(1B): 228-37. Tuc-
      son: University of Arizona Press, 1958.

1456 _____, and Oshida, I. "Use of Solar Energy for
      Water Heating." United Nations Conference on New
      Sources of Energy, E35-Gr-S13, Rome, 1961. New
      York: United Nations, 1964.

1457 _____, et al. "Construction of the Solar Radiation
      in a Solar Furnace." Solar Energy 1(4): 14-18,
      1957.

1458 Hoberman, Stuart. Solar Cell and Photocell Experi-
      menters Guide. Indianapolis, IN: Howard W. Sams,
      1964. $3. 25(pap).

1459 Hodgins, J. W. and Hoffman, T. W. "The Storage
      and Transfer of Low Potential Heat." Canadian
      Journal of Technology 33, p. 293, 1955.

1460 Hoke, John. First Book of Solar Energy. New York:
      Watts, Franklin, Inc., 1968. Illus. PLB $3. 75.
      (Gr. 4-6).

1461 _____. Solar Energy. New York: Watts, Franklin, Inc., 1968. 83pp. Illus.

1462 Holberman, Stuart. Solar Cell and Photocell Experimenters Guide. Indianapolis, IN: Howard W. Sams, 1965. 128pp. Illus., plans.

1463 Holdren, John. "Defusing Old Smokey by Plugging into Nature." Sierra Club Bulletin 56(8): 24-28, Sept. 1971. Drawing.

1464 Hollaender, A. Radiation Biology. Vols. 1-3. New York: McGraw-Hill, 1954 and 1955.

1465 Hollands, K. G. T. "A Concentrator for Thin-Film Solar Cells." Solar Energy 13: 149, 1971.

1466 _____. "Directional Selectivity, Emittance, and Absorbing Properties of Vee Corrugated Specular Surfaces." Solar Energy 7(3): 108-116, July-Sept. 1963.

1467 _____. "Honeycomb Devices in Flat-Plate Solar Collectors." Solar Energy 9: 159+, 1965.

1468 "Home Sweet Solar Home." UNESCO Courier 27(1): 26-27, Jan. 1974. Photos, diagr.

1469 Homer, Lloyd. NSF/NOAA Solar Energy Data Workshop. Washington, DC, Nov. 1973.

1470 Horowitz, Harold. "The Heating and Cooling of Buildings with Solar Energy." Papers presented at the Symposium on Solar Energy Applications at the ASHRAE Annual Meeting, June 23-27, 1974, Montreal, Canada. pp. 45-52. Graphs, tables, photos.

1471 Hottel, Hoyt C., Heywood, H., Willier, A., Lof, G. O. G., Telkes, M., and Bliss, R. W., Jr. "Panel on Solar House Heating." Proceedings, World Symposium on Applied Solar Energy, Phoenix, Arizona, 1955. Menlo Park, CA: Stanford Research Institute, 1956.

1472 _____, and Howard, Jack B. "An Agenda for Energy." Technology Review 74(4): 38ff (11pp.), Jan. 1972. Charts, graphs.

1473 _____, and Sarofim, A. F. Radiative Transfer. New York: McGraw-Hill, 1967.

1474 _____, and Unger, T. A. "The Properties of a Copper Oxide-Aluminum Selective Black Surface Absorber of Solar Energy." Solar Energy 3(3): 10-15, 1959.

1475 _____, and Whillier, A. "Evaluation of Solar Collector Performance." Transactions of the Conference on Use of Solar Energy: The Scientific Basis 2: 74-104. Tucson: University of Arizona Press, 1958.

1476 _____, and Woertz, B. B. "Performance of Flat-Plate Solar-Heat Exchangers." Transactions of the ASME 14: 91, 1942.

1477 _____, and _____. "The Performance of Plate Collectors." Transactions of the American Society of Mechanical Engineers 64: 91-104, 1942.

1478 "Hot Spot in the Pyrenees: Solar Ovens for Thermal Power." Science News 95(9): 211, March 1, 1969.

1479 Hovel, H. J. and Woodall, J. M. "Ga$_1$ - Al$_x$As-GaAs P-P-N Heterojunction Solar Cells." Journal of the Electrochemical Society 120(9): 1246-1252, Sept. 1973. Refs., illus., diagr.

1480 "How Practical Is Solar Power?" U.S. News & World Report 75(26): 62, Dec. 24, 1973. Photo., drawing.

1481 "How to Build and Use a Solar Oven." The Mother Earth News #25 pp. 26-28, Jan. 1974.

1482 "How To Get Sun Power for New York." Business Week (2123): 128, May 9, 1970. Photo.

1483 Howe, E. D. "Solar Distillation." Transactions of the Conference on the Use of Solar Energy: The Scientific Basis 3(2): 159-68. Tucson: University of Arizona Press, 1958.

1484 _____. "Solar Distillation Research at the University of California." United Nations Conference on

New Sources of Energy, E35-529, Rome, 1961.
New York: United Nations, 1964.

1485 Hsu, S. T. and Leo, B. S. "A Simple Reaction Turbine as a Solar Engine." Solar Energy 2(3-4): 7-11, 1958. Also in Solar Energy 4(2): 16-20, 1960.

1486 Hubbert, M. King. Energy Resources: A Report to the Committee on Natural Resources of the National Academy of Sciences--National Research Council. Washington, DC: NAS--National Research Council, 1962. 153pp.

1487 _____. "The Energy Resources of the Earth." Scientific American 224(3): 60ff (11pp.), Sept. 1971. Charts, diagrs. photo.

1488 _____. "Survey of World Energy Resources." Canadian Mining & Metallurgy Bulletin 66(735): 37ff (17pp.), July 1973. Diagrs., graphs, tables.

1489 Hudac, P. P. and Sonnenfeld, P. "Hot Brines on Los Roques, Venezuela." Science 185(4149): 440-2, Aug. 2, 1974. Refs., Illus.

1490 Hudock, R. P. "Harnessing the Sun." Astronautics & Aeronautics 10(7): 6-9, July 1972.

1491 "Huge Solar Array for Space Station." Space World H-3-87: 50, March 1971. Illus.

1492 Hughes, Patrick. "Building to Save Energy." NOAA 4(2): 14ff (4pp.), April 1974. Drawings, map.

1493 Hukuo, N. and Mii, H. "Design Problems of a Solar Furnace." Solar Energy 1: 108-14, 1957.

1494 Hummel, R. "Power as a By-Product of Competitive Solar Distillation." United Nations Conference on New Sources of Energy, E35-S15, Rome, 1961. New York: United Nations, 1964.

1495 Hunter, G. S. "Cell Studies Spur Solar Power Advances." Aviation Week & Space Technology 87 (9): 94-95+, Aug. 28, 1967. graphs, photo.

1496 _____. "Requirements for Solar Arrays Spurring

New Techniques. " Aviation Week & Space Technology 87(7): 72-7+, Aug. 14, 1967. Photos.,
diagr., drawing.

1497 IEEE Photovoltaic Specialists Conference (9th), Silver
Spring, MD: May 2-4, 1972. New York: IEEE,
Conference Record, 1972. 388pp.

1498 Iles, Peter A. Electroformed Aluminum Solar Cell
Contacts (Final Report). El Monte, CA: Globe-
Union, Inc., 1973. 44pp. NTIS #AD-764 816/5.

1499 "Improved Solar Cells of Cadmium Telluride. " Electronic Engineering 44: 20, March 1972.

1500 "Improving Sophisticated Energy Sources May Answer
Power Problems. " Commerce Today 3(9): 4-9,
Feb. 5, 1973. Drawings, photos.

1501 In Committee: National Science Foundation. " Congressional Quarterly 32(17): 1047-48, April 27, 1974.

1502 "In Committee: Solar Energy. " Congressional Quarterly 32(12): 764, March 23, 1974.

1503 "In the News: Solar Energy as a Power Source. "
Space World E4-52: 39, April 1968.

1504 "India Uses Sun as an Ally. " World Health p. 12-18,
Aug. -Sept. 1971.

1505 Indian National Scientific Documentation Centre. Utilization of Solar Energy: List of References compiled
by INSDOC for National Institute of Sciences of
India and UNESCO Symposium on Solar Energy and
Wind Power. Delhi: INSDOC, 1954. Refs. 72-
188808.

1506 "Industry Finances Solar Prototype. " Architectural
Record 156(5): 34, Oct. 1974. Photo.

1507 "Instruments et References Heliolux: Une Application,
Les Abaques Solaires du C. S. T. B. " L'Architecture
d'Aujourd'hui (167): 105-106, May 1973. Illus.,
diagrs.

1508 "An International Solar Energy Development Decade. "

Bulletin of the Atomic Scientists 18(5): 31ff (4pp.), May 1972.

1509 The Invisible Family Test House. St. Paul, MN: Wood Conversion Co., 1961.

1510 Irvine, T. F., Jr., Hartnett, J. P., and Eckert, E. R. G. "Solar Collector Surfaces with Wavelength Selective Radiation Characteristics." Solar Energy 2(3-4): 12-16, 1958.

1511 Irving, B. C. and Miller, E. Thermoelectric Materials and Devices. New York: Reinhold, 1960.

1512 "It's All Done with Mirrors: Archimedes' Use of Solar Power as an Effective Weapon of War." Science Digest 75(3): 18, March 1974.

1513 ITT Research Institute. Investigation of Solar Energy for Ice Melting (Final Report, June 1, 1949). Chicago: ITT Research Institute, 1949. 54 leaves. Illus.

1513a Jarvinen, P. O. "Solar Energy As a Power Source." Machine Design 46(21): 142, Sept. 5, 1974. Graph.

1514 Joffe, A. F. Semiconductor Thermoelements and Thermoelectric Cooling. London: Info-Search, 1957. (Trans. from Russian).

1515 Johnsen, K. "NASA Plans Expanded Program to Exploit Space Solar Power." Aviation Week & Space Technology 96(13): 21, March 27, 1972.

1516 Johnson, F. S. "The Solar Constant." Journal of the Royal Meterological Society 11: 431, 1954.

1517 Johnson, Tim. "The Sunshine Spreaders." New Scientist 58(845): 337-340, May 10, 1973.

1518 Johnston, Phillip A. Laboratory Experiments on the Performance of Silicon Solar Cells at High Solar Intensities and Temperatures NASA. For Sale by Office of Technical Services, Department of Commerce, 1965. 24pp. Illus.

1519 Jones, Stacey V. "Electricity Direct from the Sun."

Science Digest 72(2): 81, August 1972. Photo.

1520 _____. "Now ... Solar Air Conditioning." Science Digest 76(3): 87, Sept. 1974. Photo.

1521 _____. "Satellites Make Electricity from Sunlight." Science Digest 75(4): 70-71, April 1974. Photo., drawing.

1522 _____. "Way of Producing and Storing Cold Patented." New York Times p. 35, May 31, 1974.

1523 Jordan, Richard C. "Conversion of Solar to Mechanical Energy." Introduction to the Utilization of Solar Energy. (Zarem, A. M. and Erway, D. D., eds.) pp. 125-152. New York: McGraw-Hill, 1963.

1524 _____, ed. Low Temperature Engineering Application of Solar Energy. New York: American Society of Heating, Refrigeration, and Air Conditioning Engineers, 1967. 78pp. Illus., refs.

1525 _____. "Mechanical Energy from Solar Energy." Proceedings of the World Symposium on Applied Solar Energy: Phoenix, Arizona: Menlo Park, CA: Stanford Research Institute, 1956.

1526 _____, and Ibele, W. E. "Mechanical Energy from Solar Energy." Proceedings of World Symposium on Applied Solar Energy, Phoenix, Arizona, 1955.

1527 _____, and Threlkeld, J. L. "Availability and Utilization of Solar Energy." ASH & VE Transactions 60: 193+, 1954.

1528 Jose, P. D. "Flux Through the Focal Spot of a Solar Furnace." Solar Energy 1(4): 19-22, 1957.

1529 Justi, E. W., et al. "Investigations on CdTe Thin Film Solar Cells." Energy Conversions 13(2): 53-56, April 1973. Refs.

1530 Kagan, J. "Solar Energy: How Practical?" McCalls (11): 50-51, Aug. 1974. Illus.

1531 Kagan, M. B., Landsman, A. P., Kholev, B. A. A Study of the Electrical Characteristics of GaAs p-n

Junctions Used as Solar Cells. Trans. by Dr. Toms. Farnborough, England: Ministry of Technology, 1968. 10pp. Illus. (Royal Aircraft Establishment Library Trans. #1287).

1532 _____, and Lyubastievskaya, T. L. "Investigation of Photoelectric Characteristics of Gallium Arsenide Solar Cells Over a Wide Range of Change in Light Flux." Geliotekhnika (2): 12-21, 1971. (Trans. by Army Foreign Science and Technology Center Charlottesville, VA.) NTIS #AD-756 228.

1533 Kakabaev, A., and Golaev, M. "Experimental Study of the Thermal Output of Some Simple Solar Heater Designs." Geliotekhnika 7(2): 41-46, 1971. (Russian).

1534 Kane, J. T., ed. "Solar Energy: An Idea Whose Time Has Come and Gone and Come Again." Professional Engineer 43(10): 14-16, Oct. 1973.

1535 Kapur, J. C. "A Report on the Utilization of Solar Energy for Refrigerator and Air-Conditioning Applications." Solar Energy 4: 39-47, 1960.

1536 _____. "Socio-economic Considerations in the Utilization of Solar Energy in Under-Developed Areas." United Nations Conference on New Sources of Energy, Gen. 8, Rome, 1961. New York: United Nations, 1964.

1537 Kastens, M. "Economics of Solar Energy." Introduction to the Utilization of Solar Energy. (Zarem, A. M. and Erway, D. D., eds.) New York: McGraw-Hill, 1963.

1538 Kato, Yukio and Usami, Akira. "Effects of Boron Density on Radiation Resistance of Copper-Contaminated n/p Type Silicon Solar Cells." Electronics Letters 9(2): 18-19, Jan. 25, 1973.

1539 Katz, K. "Thermoelectric Generators for the Conversion of Solar Energy to Produce Electrical and Mechanical Power." United Nations Conference on New Sources of Energy, E35-S12, Rome, 1961.

1540 Kauffman, K. and Gruntfest, I. "Congruently Melting

Materials for Thermal Energy Storage. " Report
NCEMP-20 of the University of Pennsylvania National Center for Energy Management and Power,
to NSF, 1973.

1541 Kay, J. and Welsh, J. A. Direct Conversion of Heat
to Electricity. New York: Wiley, 1960.

1542 Kays, W. M. Convective Heat and Mass Transfer.
New York: McGraw-Hill, 1966.

1543 "Keep It In or Let It Out: Buildings That Save a Watt
and More. " Progressive Architecture 52: 104-11,
Oct. 1971. Illus. , plans, diagrs.

1543a Kelsey, P. "Congressional Hearing Airs House Bill
on Solar Energy Incentives. " Air Conditioning,
Heating and Refrigeration News 133(36): 3+, Sept.
9, 1974.

1544 Kent, M. F. "Power Generation Options for the
Eighties and Nineties. " Public Utilities Fortnightly
90(4): 17ff (5-1/2pp.), Aug. 17, 1972. Drawings.

1545 Kenward, Michael. "Here Comes the Sun. " New Scientist 59(854): 71-73, July 12, 1973.

1546 Keyerleber, K. "Clean Power From the Sun. " Current (158): 7-12, Jan. 1974. Also in Nation 217
(14): 429-32, Oct. 29, 1973.

1547 Khanna, M. L. "Design Data for Heating Air by Means
of Heat Exchanges with Reservoir, Under Free Convection Conditions, For Utilization of Solar Energy."
Solar Energy 12: 447-56, 1969.

1548 _____, and Mathur, K. N. "Experiments on Demineralization of Water in North India. " United
Nations Conference on New Sources of Energy, E35-
S115, Rome, 1961. New York: United Nations,
1964.

1549 Klaff, Jerome L. "Power Policies. " Public Utilities
Fortnightly 92(6): 31-34, Sept. 13, 1973. Drawing.

1550 Klass, P. J. "Gains Cited in Thin-film Solar Cell
Efforts. " Aviation Week & Space Technology 89(9):
74-76, Aug. 26, 1968. Photos. , diagr.

1551 _____. "Research Hikes Satcom Efficiency." Aviation Week & Space Technology 99(11): 39-41, Sept. 10, 1973. Photos, diagr.

1552 Klein, S. A. "The Effects of Thermal Capacitance Upon the Performance of Flat-Plate Solar Collectors." M. S. Thesis in Chemical Engineering, Madison: University of Wisconsin, 1973.

1553 _____, Cooper, P. I., Beckman, W. A., and Duffie, J. A. "TRNSYS, A Transient Simulation Program." Madison, University of Wisconsin, Engineering Experiment Station Report #38, 1974.

1554 Kobayashi, M. "A Method of Obtaining Water in Arid Lands." Solar Energy 7: 93-99, 1963.

1555 _____. "Thermoelectric Generator." United Nations Conference on New Sources of Energy, E35-S10, Rome, 1961. New York: United Nations, 1964.

1556 _____. "Utilization of Silicon Solar Batteries." United Nations Conference on New Sources of Energy, E35-S11, Rome, 1961. New York: United Nations, 1964.

1557 Kochar, A. S. "India Uses the Sun as an Ally." World Health p. 12-18, Aug. -Sept. 1971. Photos.

1558 Kok, B. "The Yield of Sunlight Conversion by Chlorella." Transactions of the Conference on the Use of Solar Energy: The Scientific Basis 4: 48-54. Tucson: University of Arizona Press, 1958.

1559 Kokoropoulos, P. and Evans, M. "Infrared Spectral Emissivities of Cobalt Oxide and Nickle Oxide Films." Solar Energy 8: 69-73. 1964.

1560 _____, Salam, E., and Daniels, F. "Selective Radiation Coatings, Preparation & High Temperature Stability." Solar Energy 3(4): 19-23, 1959.

1561 Kolodkine, Paul. La Conquete de l'Energie Solaire. Pref. de Felix Frombe. Paris: Editions La Farandole, 1960. 181pp. Illus.

1562 Koltun, M. M. "Selective Surfaces and Coating in

Solar Radiation Engineering. " Geliotekhnika (5): 70-80, 1971. Trans. by Scientific Translation Service, Santa Barbara, CA.

1563 Korolev, M. "Space Electric Power Plants, Part 2." Pravda (161): 3, June 10, 1971. Trans. by Army Foreign Science and Technology Center, Charlottesville, VA. NTIS #AD-756 039.

1564 Kozyrev, B. P. and Vershinin, O. E. "Determination of Spectral Coefficients of Diffuse Reflection of Infrared Radiation from Blackened Surfaces. " Optics and Spectroscopy 6: 345, 1959.

1565 Krpichev, M. V. and Barim, V. A. "Exploitation of Sun's Rays. " Privoda 43: 45, 1954.

1566 Kubaschewski, O. and Evans, E. L. Metallurgical Thermo-Chemistry, 3rd ed. Elmsford, NY: Pergamon Press, 1958.

1567 Kupperman, Robert H. "Energy Supply Not Keeping Pace with Demand. " Mining Congress Journal 59 (3): 25-29, March 1973.

1568 Kurz, J. L. "A Combination Boiler-reaction Turbine as a Solar Engine. " Special Project in Mechanical Engineering 151 with Professor S. T. Hsu, University of Wisconsin, 1962.

1569 Kusuda, T. "Solar Water Heating in Japan. " Proceedings of the Solar Heating and Cooling for Buildings Workshop, NSF/RANN-73-004, pp. 141-47, July 1973.

1570 Laird, L. "What About Solar Energy?" House and Home 45(2): 87, Feb. 1974.

1571 LaMer, V. K. , ed. Retardation of Evaporation by Mono-layers. New York: Academic Press, 1962.

1571a Landa, Henry C. Solar Energy Handbook, 2nd ed. Milwaukee: FICOA, 1974. Illus. $9. 00 (pap). (Film Instruction Co. of America).

1572 Landry, B. A. "The Storage of Heat & Electricity. " Proceedings of the National Academy of Science 47:

1290-95, 1961. Also in Solar Energy (Special Issue) 5: 46-51, Sept. 1961.

1573   Lanzerotti, L. J. "Solar Proton Radiation Damage of Solar Cells at Synchronous Altitudes." Journal of Spacecraft and Rockets 6: 183-84, Nov. 24, 1969.

1574   Lappala, R. and Bjorksten, J. "Development of Plastic Solar Stills for Use in the Large-Scale, Low-Cost Demineralization of Saline Waters by Solar Evaporation." Transactions of the Conference on the Use of Solar Energy: The Scientific Basis 3(2): 99-107. Tucson: University of Arizona Press, 1958.

1575   Laszlo, T. S. Image Furnace Techniques. New York: Wiley-Interscience, 1965.

1576   _____. "Measurement and Applications of High Heat Fluxes in a Solar Furnace." Solar Energy 6(2): 69-73, 1962.

1577   _____. "New Techniques and Possibilities in Solar Furnaces, #16." United Nations Conference on New Sources of Energy, E35-S, Rome, 1961. New York: United Nations, 1964.

1578   _____, de Dufour, W. F., and Erdell, J. A. "A Guiding System for Solar Furnaces." Transactions of the Conference on the Use of Solar Energy: The Scientific Basis 2(1B): 222-27, Tucson: University of Arizona Press, 1958.

1579   Lauck, F. W., Busch, C. W., Uyehara, O. A., and Myers, P. S. "Portable Power from Non-Portable Energy Sources." Society of Automotive Engineers, Paper 608A, National Power Plant Meeting, Philadelphia, 1962.

1580   _____, Myers, P. S., Uyehara, O. A., and Glander, H. "Mathematical Model of a House and Solar-Gas Absorption Cooling and Heating System." ASHRAE Transactions 71: 273, 1965.

1581   Lavi, A. and Zener, C. "Plumbing the Ocean Depths: A New Source of Power." IEEE Spectrum 10(10): 22-27, Oct. 1973. Refs., diagrs.

Solar Energy / 69

1582 Lawand, T. A. "Technical Evaluation of a Large-
      Scale Solar Distillation Plant. " Journal of Engineer-
      ing for Power 92: 95-102, April 1970. Refs.,
      illus. , diagrs.

1583 Lee, D. G. "Tapping the Sun's Energy. " National
      Wildlife 12(5): 18-21, Aug. 1974. Illus.

1584 Lejeune, G. and Savornin, J. "Recovering Water
      Vapor from the Atmosphere. " Transactions of the
      Conference on the Use of Solar Energy: The Scien-
      tific Basis 3: 138-40. Tucson: University of Ari-
      zona Press, 1958.

1585 Le-Phat-Vinh, A. "Etude sur les Concentrations En-
      ergetiques Données pars les Miroirs Paraboliques
      de très Grand Surface. " United Nations Conference
      on New Sources of Energy, E35-S, Rome, 1961.
      New York: United Nations, 1964.

1586 Levinson, J. "Using the Sun to Brighten the Crisis."
      AIA Journal 61: 40-43, Feb. 1974. Illus., plans,
      tables, diagrs.

1587 Lidorenko, N. S. , Nabiullin, F. Ka, Landsman, A. P.,
      Tarnizhevski, B. V. , Gertsik, E. M. , and Shul'-
      meister, L. F. "An Experimental Solar Power
      Plant. " Geliotekhnika 1(3): 5-9, 1965. (Russian).

1588 Liebert, C. H. and Hibbard, R. R. "Performance of
      Spectrally Selective Collector Surfaces in a Solar
      Driven Carnot Space-Power System. " Solar Energy
      6(3): 84-88, 1962.

1589 Liebhafsky, H. A. and Cairns, E. J. "The Hydro-
      carbon Fuel Cell. " Chemical Engineering Progress
      59(10): 35-37, 1962.

1590 "Lighting Some Candles. " Architectural Forum 139
      (1): 89ff (10pp.), July-Aug. 1973. Drawing, photos.

1591 Lindmayer, J. "Theoretical and Practical Fill Fac-
      tors in Solar Cells. " COMSAT Technical Review
      2(1): 105-121, Spring 1972. Refs.

1592 _____, and Revesz, A. G. "Electronic Process in
      CUS-CdS Photovoltaic Cells. " Solid State Elec-
      tronics 14: 647-54, Aug. 1971. Diagrs.

1593   Litzinski, L.   Direct Energy Conversion in the
       U. S. S. R.:   Soviet Solar Concentrators.   Compre-
       hensive Report.   Washington, DC: U. S. Library of
       Congress Aerospace Technology Division, 1966.
       138 leaves.   Illus., refs.   67-61890.

1594   Liu, B. Y. H., and Jordan, R. C.   "Availability of
       Solar Energy for Flat-Plate Solar Heat Collectors."
       Low Temperature Engineering Applications of Solar
       Energy, New York, American Society of Heating,
       Refrigeration, and Air Conditioning Engineers, 1967.

1595   _____, and _____.   "Performance and Evaluation
       of Concentrating Solar Collectors for Power Gen-
       eration."   Journal Engineering Power 87: 1-12,
       1965.

1596   _____, and _____.   "Predicting Long Term Aver-
       age Performances of Flat Plate Solar Energy Col-
       lectors."   Solar Energy 7: 53, 1963.

1597   _____, and _____.   "The Interrelationship and
       Characteristic Distribution of Direct, Diffuse, and
       Total Solar Radiation."   Solar Energy 4(3): 1-19,
       1960; 7: 71-74, 1963.

1598   Livingston, R.   "The Photochemistry of Chlorophyll in
       Vitro."   Transactions, Conference on the Use of
       Solar Energy: The Scientific Basis 4: 138-44.
       Tucson: University of Arizona Press, 1958.

1599   _____.   "The Role of the Triplet State in the Photo-
       chemical Auto-Oxidation of Aryl Hydrocarbons."
       Photochemistry in the Liquid and Solid States.
       (Heidt, L. J., et al.)   New York: Wiley, 1960.
       pp. 76-82.

1600   Lodding, W.   "Desert Lakes and Solar Power."   En-
       vironment 16(1): 43-44, Jan. 1974.   Refs.

1601   Löf, George O. G.   "Application of Theoretical Prin-
       ciples in Improving the Performance of Basin-Type
       Solar Distillers."   United Nations Conference on
       New Sources of Energy, E35-S77, Rome, 1961.
       New York: United Nations, 1964.

1602   _____.   "Cooling with Solar Energy."   Proceedings

of the World Symposium on Applied Solar Energy, Phoenix, Arizona, 1955. Menlo Park, CA: Stanford Research Institute, 1956. pp. 171-89.

1603 _____. "Fundamental Problems in Solar Distillation." Proceedings of the National Academy of Science 47: 1279-89, 1961; Also in Solar Energy (Special Issue) 5: 35-45, Sept. 1961.

1604 _____. "House Heating and Cooling with Solar Energy." Solar Energy Research (Daniels, F. and Duffie, J. A., eds.) Madison: University of Wisconsin Press, 1955. pp. 33-55.

1605 _____. "Profits in Solar Energy." Solar Energy 4(2): 9-16, 1960.

1606 _____. "Solar Cooling Design and Cost Study." Proceedings of the NSF/RANN Solar Heating and Cooling for Buildings Workshop, NSF/RANN-73-004, pp. 119-25, July 1973.

1607 _____. "Solar Energy: An Infinite Source of Clean Energy." Annals of the American Academy of Political and Social Science 410: 52-64, Nov. 1973.

1608 _____. "Solar Energy for the Drying of Solids." Solar Energy 6: 122-28, 1962.

1609 _____. Article by Löf in Solar Energy Research (Daniels, F. and Duffie, J. A., eds.) Madison: University of Wisconsin Press, 1955. pp. 43-45.

1610 _____. "Solar Heating and Cooling of Buildings-- Background and Economic Factors." Energy Environment Productivity, Proceedings of the First Symposium on RANN: Research Applied to National Needs, Washington, D. C., November 18-20, 1973. Washington: National Science Foundation. pp. 46-49. Diagr., graph, tables.

1611 _____. "The Use of Solar Energy for Cooking." United Nations Conference on New Sources of Energy, E35-Gr-S16, Rome, 1961. New York: United Nations, 1964. Also in Solar Energy 7: 125-33, 1953.

1612 _____. "The Use of Solar Energy for Space Heating, General Report." United Nations Conference on New Sources of Energy, E35-Gr-S14, Rome, 1961. New York: United Nations, 1964.

1613 _____, and Close, D. J. Low Temperature Engineering Application of Solar Energy. New York: ASHRAE, 1967.

1614 _____, and Duffie, J. A. "Optimization of Focusing Solar-Collector Design." Journal of Engineering for Power 85A: 221-28, July 1963.

1615 _____, _____, and Smith, C. O. "World Distribution of Solar Energy." Solar Energy 10(1): 27-32, 1966.

1616 _____, _____, and _____. "World Distribution of Solar Radiation." Engineering Experiment Station Report 21. Madison: University of Wisconsin, July 1966a.

1617 _____, El-Wakil, M., and Chiou, J. "Design and Performance of Domestic Heating System Employing Solar-Heated Air, the Colorado Solar House." United Nations Conference on New Sources of Energy, E35-S114, Rome, 1961. New York: United Nations, 1964.

1618 _____, and Fester, D. A. "Design and Performance of Folding Umbrella-Type Solar Cooker." United Nations Conference on New Sources of Energy, E35-S100, Rome, 1961. New York: United Nations, 1964.

1619 _____, _____, and Duffie, J. A. "Energy Balances on a Parabolic Cylindrical Solar Collector." Transactions of American Society of Mechanical Engineers rp. 24-32, 1932.

1620 _____, and Hawley, R. W. "Unsteady State Heat Transfer Between Air and Loose Solids." Industrial Engineering Chemistry 40: 1061, 1948.

1621 _____, and Tybout, R. A. "Cost of House Heating with Solar Energy." Solar Energy 14: 253-78, 1973.

1622 _____, and _____. "The Design and Cost of
Optimized Systems for Cooling Dwellings by Solar
Energy." Paper presented at the International Solar
Energy Society Congress, Paris, 1973.

1623 _____, and Ward, Dan S. "Solar Heating and Cooling: Untapped Energy Put to Use." Civil Engineer-
ing--ASCE 43(9): 88-92, Sept. 1973. Diagrs.,
photos.

1624 _____, et al. "Energy Balances on a Parabolic
Cylindrical Solar Collector." Transactions of the
Society of Mechanical Engineers, p. 24-32, 1962.

1625 Loferski, J. J. "Large Scale Solar Power via the
Photovoltaic Effect." Mechanical Engineering 95
(12): 28-32, Dec. 1973. Diagrs.

1626 _____. "Some Problems Associated with Large
Scale Production of Electrical Power from Solar
Energy via the Photovoltaic Effect." American
Society of Mechanical Engineers, Papers N. 72-WA/
Sol-4 for Meeting Nov. 26, 1972. 4pp. Refs.

1627 _____. "Theoretical Considerations Governing the
Choice of the Optimum Semi-conductor for Photo-
voltaic Solar Energy Conversion." Transactions of
the Conference on the Use of Solar Energy: The
Scientific Basis 5: 65-98. Tucson: University of
Arizona Press, 1958.

1628 Löh, E., Hiester, N. K., and Tietz, T. E. "Heat
Flux Measurements at the Sun Image of the Cali-
fornia Institute of Technology Lens-Type Solar
Furnace." Solar Energy 1(4): 23-26, 1957.

1629 Lorsch, Harold G. "Performance of Flat Plate Collectors." Proceedings of the NSF/RANN Solar Heating and Cooling for Buildings Workshop, Washington,
DC: NSF/RANN-73-004, pp. 1-14, July 1973.

1630 _____. "Solar Heating/Cooling Projects at the
University of Pennsylvania." Proceedings of the
NSF/RANN Solar Heating and Cooling for Buildings
Workshop, NSF/RANN-73-005, pp. 194-210, July
1973.

74 / Alternate Sources of Energy

1631 _____, and Niyogi, B. "Influence of Azimuthal Orientation on Collectible Energy in Vertical Solar Collector Building Walls." Report NSF/RANN/SE /GI 27976/TR72/18, to NSF from University of Pennsylvania, Aug. 1971.

1632 Lovins, Amory B. "The Case Against the Fast Breeder Reactor." Bulletin of Atomic Scientists 29(3): 29ff (7pp.), March 1973. Refs.

1633 _____. "Energy in the Real World." Not Man Apart. 3(12): 8ff (7pp.), Dec.-Jan. 1974. Drawings, table.

1634 "Lumidue Architecture: La Cité Solaire, la Ville Souterraine." L'Architecture d'Aujourd'hui (155): xlviii, April 1971. Diagrs.

1635 McAdams, W. C. Heat Transmission, 3rd ed. New York: McGraw-Hill, 1954.

1636 McAllan, J. V. "Storage of High Grade Energy." Paper presented at Australia/U.S. Solar Energy Workshop at Sydney and Melbourne, 1974.

1637 McCallum, John and Faust, Charles L. "New Frontiers in Energy Storage." Battelle Research Outlook 4(1): 26-30, 1972. Chart, photo.

1638 McCaull, Julian. "Oil Drum Technology." Environment 15(7): 13-16, Sept. 1973. Photos.

1639 McCormack, [U.S. Rep.] Mike. "Energy Crisis--An In-Depth View." Chemical & Engineering News 51(14): 1-3, April 2, 1973.

1640 _____. "We Must Explore New Methods of Energy Conversion." Public Power 31(1): 20-24, Jan. Feb. 1973. Diagr.

1641 McCully, C. "The Chemical Conversion of Solar Energy." United Nations Conference on New Sources of Energy, E35-Gen. 6, Rome, 1961. New York: United Nations, 1964.

1642 MacDonald, D. K. C. Thermoelectricity: An Introduction to the Principles. New York: Wiley, 1967.

1643 McDow, John. Factors Affecting the Design of a Solar
Energy Storage Unit. Michigan State University.
Dissertation #59-5597.

1644 McElroy, T. "Solar Heating Seen Ready Commercially." Commercial and Financial Chronicle 218:
1485, Nov. 19, 1973.

1645 McGough, E. "Extend Your Swimming Season with
Sun Power." Mechanix Illustrated 67(520): 98-99,
Sept. 1972. Drawing, photos.

1646 McGowan, J. G., Connell, J. W., Ambs, L. L., and
Goss, W. P. "Conceptual Design of a Rankine
Cycle Powered by the Ocean Thermal Difference."
Proceedings of the Intersociety Energy Conversion
Engineering Conference (IECED), paper no. 739120,
pp. 420-27, Aug. 1973.

1647 MacKillop, Andrew. "Living Off the Sun." Ecologist
3(7): 260-265, July 1973.

1648 McManus, George J. "First There Was Steam, Then
Nuclear, and Now Sea Solar Power." Iron Age
212(3): 34, July 19, 1973. Diagr., photo.

1649 McNally, P. J. Development of GAAS Solar Cells:
Quarterly Technical Report 31 Jan.-30 April 1972.
Burlington, MA: Ion Physics Corp., 1972. 30pp.
NTIS #N73-11041.

1650 _____. Development of GAAS Solar Cells: Quarterly
Technical Report 1 May-31 July 1972. Burlington,
MA: Ion Physics Corp., 1972. 21pp. NTIS #N73-
11042.

1651 MacNevin, W. C. "The Trapping of Solar Energy."
Ohio Journal of Science 53(5): 257-319, 1953.

1652 McQuigg, James D. and Decker, Wayne L. Solar Energy: A Summary of Records at Columbia, Missouri. Columbia: MO: Missouri Agricultural Experiment Station (Research Bulletin #671), 1958.
27pp. Illus., diagrs., tables.

1653 Makhijani, A. B. and Lichtenberg, A. J. "Energy
and Well-Being." Environment 14(5): 10ff (9pp.),

June 1972. Charts, drawing, photos.

1654 Malevskii, Yu. N. "Performance Reliability Calculation for a Modular Solar Thermoelectric Generator." Geliotekhnika (1): 16-20, 1971. NTIS
#AD-757 087.

1655 Malik, M. A. S. "Solar Water Heating in South
Africa." Solar Energy 12: 395-97, 1969.

1656 Mandelkorn, J. and Lamneck, J. H., Jr. "New Electric Field Effect in Silicon Solar Cells." Journal
of Applied Physics 44(10): 4785-87, Oct. 1973.
Diagrs., refs., tables.

1657 Mangelsdorf, P. C. "The World's Principal Food
Plants as Converters of Solar Energy." Transactions of the Conference on the Use of Solar Energy: The Scientific Basis 4: 122-29. Tucson:
University of Arizona Press, 1958.

1658 Marcus, R. J. "Chemical Conversion of Solar Energy." Science 123(3193): 399-405, March 9, 1956.
Graphs, tables, diagrs.

1659 _____. The Hill Reaction as a Model for Chemical
Conversion of Solar Energy, Publ. PB 161, 462.
Washington, DC.

1660 _____. "Survey of Electron Transfer Spectra of
Inorganic Ions." Solar Energy 4(4): 20-23, 1960.

1661 _____, and Wohlers, H. C. "Chemical Conversion
and Storage of Concentrated Solar Energy." Proceedings of the UN Conference on New Sources of
Energy 1, 1964. See also: Solar Energy 5: 121,
1961; 5: 44, 1961; 4(2): 1, 1960.

1662 _____, and _____. "Chemical Syntheses in the
Solar Furnace." United Nations Conference on New
Sources of Energy, E35-S, Rome, 1961. New
York: United Nations, 1964.

1663 _____, and _____. "Photolysis of Nitrosyl Chloride." Solar Energy 4(2): 1-8, 1960.

1664 Marine, G. "Alternative Energy: Here Comes the

Sun. " Ramparts 12(8): 33-37, March 1974. Illus.

1665 "Marketing Approaches to be Studies to Develop Solar
Climate Controls. " AIA Journal 60: 88, Sept. 1973.

1666 Markus, T. "Solar Energy and Building Design. "
Architectural Review (135): 456+; (136): 68+, June
and July, 1964. Illus.

1667 Marshall, E. "Power from the Sun. " New Republic
169(24): 15-17, Dec. 15, 1973.

1668 Martin, D. C. and Bell, R. "The Use of Optical In-
terference to Obtain Selective Energy Absorption. "
Proceedings of the Conference on Coatings for the
Aerospace Environment, WADD-TR-60-TB, Wright
Air Development Division, Dayton, Ohio, Nov. 1960.

1669 Martinek, Frank. "Heat-Transfer Process in Solar
Energy Storage Systems for Orbital Applications. "
Journal of Spacecraft and Rockets 7: 1032-37, Sept.
1970. Refs., diagrs.

1670 _____. Investigation of Heat Transfer Process in
Solar Energy Storage Systems for Space Application. University of Cincinnati. Dissertation.
#66-11484.

1671 Maryland University Bureau of Business and Economic
Research. Solar and Atomic Energy: A Survey.
College Park: Maryland University Studies in Busi-
ness and Economics 12(4): 21pp.

1672 Masson, H. "Les Insolateurs à Bas Potential. "
Transaction of the Conference on the Use of Solar
Energy: The Scientific Basis 3(2): 47-66. Tucson:
University of Arizona Press, 1958.

1673 _____, and Girardier, J. P. "Solar Motors with
Flat-Plate Collectors. " Solar Energy 10: 165, 1966.

1674 Mathur, K. N. "Use of Solar Energy for Heating Purposes, Heat Storage. " United Nations Conference
on New Sources of Energy, E35-Gr-S17, Rome,
1961. New York: United Nations, 1964.

1675 _____, and Khanna, M. L. "Solar Water Heater. "

United Nations Conference on New Sources of Energy, E35-S102, Rome, 1961. New York: United
Nations, 1964.

1676 _____, _____, et al. "Domestic Solar Water
Heater." Journal of Scientific Industrial Research
(18A): 15-58, Feb. 1959.

1677 May, G. "Ein 'Sonnenhaus' für Heisse, Trockene
Gebiete." Baukunst und Werkform 15(4): 212+,
1962.

1678 Meier, R. L. "Biological Cycles in the Transforma-
tion of Solar Energy into Useful Fuels." Solar
Energy Research (Daniels, F. and Duffie, J. A.,
eds.) Madison: University of Wisconsin Press,
1955. pp. 179-84.

1679 Meinel, A. B. "A Joint United States-Mexico Solar
Power and Water Facility Project." Optical Sciences Center, Tucson: University of Arizona, April
1971.

1680 _____, and Meinel, M. P. "Energy Research and
Development." Hearings of the House Committee
on Science and Astronautics. pp. 583-85. Washington, DC: U.S. Government Printing Office, 1972.

1681 _____, and _____. "Is It Time for a New Look
at Solar Energy?" Bulletin of Atomic Scientists
27(8): 32ff (6pp.), Oct. 1971. Diagr., map, photo.

1682 _____, and _____. "Physics Looks at Solar Energy." Physics Today 25(2): 44-50, Feb. 1972.
Photo, graphs, diagrs., refs. Discussion 25(5):
15+, May 1972. Illus.

1683 _____, and _____. "Proposed Solar Plant Would
Produce Power and Water in the Desert Southwest."
IEEE Spectrum 8(6): 84, June 1971. (1)

1684 _____, and _____. "Solar Energy--The Possible
Dream?" Aware (17): 3-6, Feb. 1972. Diagr.

1685 _____, and _____. "The Solar Power Option."
Technical Report Presented at the Annual Meeting
of the American Physical Society and American

Association of Physics Teachers, San Francisco, CA, Feb. 2, 1972. 15pp.

1686 _____, and _____. "Thermal Conversion Offers Potential in Solar Energy Production. " Aware (33): 7-11, June 1973.

1687 Merlin, C. "Condensation of Atmospheric Water Vapor. " Discussion in Conference on Solar and Aeolian Energy, Sounion, Greece, 1961. New York: Plenum Press, 1964. p. 465.

1688 Merriam, Marshall E. "Materials Technology for Flat Plate Steam Generation. " Proceedings of the NSF/ RANN Solar Heating and Cooling for Buildings Workshop, NSF/RANN-73-004, pp. 17-20, July 1973.

1689 Merrill, Richard, et al. Renewable Sources: An Energy Primer. Berkeley, CA: Portola Institute, 1974. 112p. Illus. $4.

1690 Metz, William D. "Ocean Temperature Gradients: Solar Power from the Sea. " Science 180(4092): 1266-67, June 22, 1973. Map.

1691 Michel, R. "New Solar Thermoelectric Generators-- Description, Results, and Future Prospects. " United Nations Conference on New Sources of Energy, E35-S55, Rome, 1961. New York: United Nations, 1964.

1692 Middleton, A. E. , et al. Effects of Positive Ion Implantation Into Antireflection Coating of Silicon Solar Cells. Columbus, OH: Ohio State University, 1971. 150pp. NTIS #N73-19059.

1693 Mills, M. M. The Potential of Direct Solar Energy in Planning. C. P. L. Bibliography No. 476, 1973. 13pp.

1694 Mockovcizk, John, Jr. "A Systems Engineering Overview of the Satellite Power Station. " Proceedings of the 9th Intersociety Energy Conversion Engineering Conference (IECEC), paper no. 739111, pp. 712-19, Sept. 1972.

1695 Molenaar, A. "Water-Lifting Devises for Irrigation."
      New York: Food & Agricultural Organization, United
      Nations, 1956.

1696 Momota, T. and Matsurkura, Y. "The Efficiency of
      Solar Thermoelectric Generators Composed of Semi-
      conductors." Transactions of the Conference on the
      use of Solar Energy: The Scientific Basis 5: 21-30.
      Tucson: University of Arizona Press, 1958.

1697 Moon, P. "Proposed Standard Solar Radiation Curves
      for Engineering Use." Journal of the Franklin In-
      stitute 230: 583-617, 1940.

1698 Moorcraft, Colin. "Solar Energy in Housing." Arch-
      itectural Design 43: 634ff (20pp.), Oct. 1973.
      Photos, graphs.

1699 _____. "Solar Energy: Plant Power." Architec-
      tural Design 44(1): 18ff (12pp.), Diagrs., drawing,
      graph, photo, tables.

1700 Moore, J. G. and St. Amand, P. "A Guidance Sys-
      tem for a Solar Furnace." Solar Energy 1(4): 27-
      29, 1957.

1701 Morgen, R. A. "The Heat Pump." Solar Energy Re-
      search. (Daniels, F. and Duffie, J. A., eds.)
      Madison: University of Wisconsin Press, 1955. pp.
      69-70.

1702 Morikofer, W. "On the Principles of Solar Radiation
      Measuring Instruments" and "Methods for the Open-
      -Air Measurements of Caloric Radiation." Trans-
      actions of the Conference on Use of Solar Energy
      1: 60 and 63. Tucson: University of Arizona
      Press, 1958.

1703 Morrison, Clayton, A. and Farber, Erich A. "De-
      velopment and Use of Solar Insolation Data in
      Northern Latitudes for Facing Surfaces." Papers
      presented at the Symposium on Solar Energy Appli-
      cations at the ASHRAE Annual Meeting, June 23-
      27, 1974, Montreal, Canada. pp. 4-16. Graphs,
      refs., tables.

1704 Morrow, Walter E., Jr. "Solar Energy: Its Time is

Near." Technology Review 76(2): 31-43, Dec. 1973.
Diagrs., graphs, map, photos, tables.

1705 Morse, R. N. Installing Solar Water Heaters, Circular 1. Melbourne, Australia, 1959.

1706 _____. "Solar Water Heaters." Proceedings, World Symposium on Applied Solar Energy, pp. 191-200. Menlo Park, CA: Stanford Research Institute, 1956.

1707 _____. "Solar Water Heaters for Domestic and Farm Use." Commonwealth Scientific and Industrial Research Organization, Engineering Section Report ED 5, 1957.

1708 _____. "Water Heating by Solar Energy." United Nations Conference on New Sources of Energy, E35-S38, Rome, 1961. New York: United Nations, 1964.

1709 _____, and Czarnecki, J. T. "Flat Plate Solar Absorbers: The Effect on Incident Radiation of Inclination and Orientation." Report E.E. 6 of Engineering Section (Now Mechanical Engineering Division), Commonwealth Scientific and Industrial Research Organization, Melbourne, Australia, 1958.

1710 Moumoni, Abdou. "Energy Needs and Problems in the Sahelian and Sudenese Zones: Prospects of Solar Power." Ambio 2(6): 203ff (11p.), 1973. Drawing, maps, photos.

1711 Muecke, M. M. "Taming the Energy of the Sun." Product Engineering 40(14): 130, July 14, 1960. Photo.

1712 Mullin, J. P., et al. "Solar-cell-powered Electric Propulsion for Automated Space Missions." Journal of Spacecraft and Rockets 6: 1217-25, Nov. 1969. Refs., diagrs.

1713 Murray, W. D. and Landis, F. "Numerical and Machine Solutions of Transient Heat Conduction Problems Involving Melting or Freezing." Trans. ASME, J. Heat Transfer 81C: 107, 1959.

1714 Myers, J. "Algae as an Energy Converter." Pro-
      ceedings, World Symposium on Applied Solar En-
      ergy, Phoenix, Arizona, 1955, pp. 227-30. Menlo
      Park, CA: Stanford Research Institute, 1956.

1715 Nash, Barbara. "Energy from the Earth and Beyond."
      Chemistry 46(9): 6-10, Oct. 1973. Map, diagrs.,
      photo.

1716 National Center for Space Studies (France). Solar
      Cells. New York: Gordon & Breach Science Pub.,
      1971. $29.50.

1717 National Research Council, Building Research Institute.
      Solar Effects on Building Design. Washington, DC:
      Building Research Institute, 1957. 201pp.

1718 "Naturale-Artificiale Micro-clima con le Lenti Solare."
      Domus (515): 5-6, Oct. 1972. Illus., diagrs.

1719 Nebbia, G. "Present Status and Future of Solar
      Stills." United Nations Conference on New Sources
      of Energy, E35-S113, Rome, 1961. New York:
      United Nations, 1964.

1720 _____. "The Problem of Obtaining Water from the
      Air." Conference on Solar & Aeolian Energy, So-
      union, Greece, 1961. New York: Plenum Press,
      1964. pp. 33-47.

1721 Nenness, J. R. "Recommendations and Suggested
      Techniques for the Manufacture of Inexpensive Solar
      Cookers." Solar Energy 4: 22-24, 1960

1722 Neuwirth, O. S. "The Photolysis of Nitrosyl Chloride
      and the Storage of Solar Energy." Journal of Phys-
      ical Chemistry 63: 17, 1959. Also in Photochem-
      istry in the Liquid & Solid States (Heidt, L. J.,
      et al.) New York: Wiley, 1960. pp. 16-20.

1723 "New Glass Means Longer Life for Satellites." Space
      World H-10-94: 45, Oct. 1971. Illus.

1724 "New Solar Cells Have Improved Efficiencies." Elec-
      tronics and Power 18: 279, July 1972.

1725 Nice, Joan. "George Lof: Solar Scientist." High

Country News 6(7): 15-16, March 29, 1974. Photo., drawing.

1726   . "Harnessing Limitless Energy." High Country News 6(7): 1ff (3-1/3pp.), March 29, 1974. Photos., drawing.

1727 Nicoletta, C. A. Effects of Electron Irradiation and Temperature on 1 Ohm-Cm and 10 Ohm-Cm Silicon Solar Cells. Greenbelt, MD: Goddard Space Flight Center, 1973. 36pp. NTIS #N73-21083. (4)

1728   . Effects of Proton Irradiation and Temperature on 1 Ohm-Cm and 10 Ohn-Cm Silicon Solar Cells. Greenbelt, MD: NASA Goddard Space Flight Center, 1973. 42pp. NTIS #N73-16031/GA.

1729 Nilsen, Joan M. "Is Solar Energy Ready to Soar?" Chemical Energy 80(22): 24-27, Oct. 1, 1973. Diagr., photo.

1730 Nixon, Richard M. "Clean Energy." Message from the President of the United States, Document No. 92-118, June 4, 1971. 12pp.

1731 Noguchi, Tetsuo. "Recent Developments in Solar Energy Research and Application in Japan." Solar Energy 15(2): 179-87, July 1973. Refs.

1732 Norris, D. J. "Correlation of Solar Radiation with Clouds." Solar Energy 12: 107, 1968.

1732a "Now You Can Buy Solar Heating Equipment for Your Home." Popular Mechanics 206(3): 74ff (6pp.), March 1975. Photos., drawings, diagrs., buying guide.

1733 Noyes, R. M. "Kinetic Complications Associated with Photochemical Storage of Energy." Photochemistry in the Liquid & Solid States (Heidt, L. J., et al.). New York: Wiley, 1960. pp. 70-74.

1734 Noyes, W. A., Jr. and Leighton, P. A. The Photochemistry of Gases. New York: Reinhold, 1941.

1735 "NSF to Spend $2.5 million on Solar Energy Projects." Air Conditioning, Heating, and Refrigeration News

130(6): 1+, Oct. 8, 1973.

1736 Nutrition Division, F. A. O. "Report on Tests Conducted Using Telkes Oven and the Wisconsin Solar
Stove." United Nations Conference on New Sources
of Energy, E35-S116, Rome, 1961. New York:
United Nations, 1964.

1737 O'Brien, B. J. "Degradation of Apollo 11 Deployed
Instruments Because of Lunar Module Ascent Ef-
fects." Journal of Applied Physics 41(11): 4538-41,
Oct. 1970. Diagr., graphs.

1738 O'Connor, Egan. "Moratorium Politics." Not Man
Apart 3(5): 10-12, May 1973. Drawings, table.

1739 _____. "Not So Exotic: Solar Power for the Seventies." High Country News 6(1): 1ff (6pp.), Jan. 4,
1974. Photos., graphs.

1740 _____. "Solar Power for the Seventies." High
Country News 6(1): 1ff (4pp.), 1974.

1741 Oddo, S. "How Soon Can Energy from the Sun, the
Earth, and the Air Be Put to Work for You?"
House & Garden Incorporating Living for Young
Homemakers 145: 70-1+, Jan. 1974. Illus.

1742 Odum, Howard T. Environment, Power, and Society.
New York: Wiley-Interscience, 1971. Index, refs.,
pp. 1-3, 22, 50, 72, 123, and 257.

1743 Olgyay, A. "Design Criteria of Solar Heated House."
United Nations Conference on New Sources of En-
ergy, S35-S93, Rome, 1961. New York: United
Nations, 1964.

1744 _____, and Olgyay, V. Solar Control and Shading
Devices. Princeton, NJ: Princeton University,
1957. 201pp.

1745 _____, and Telkes, M. "Solar Heating for Houses."
Progressive Architecture March 1959. pp. 195-207.

1746 Oman, H. and Bishop, C. J. "A Look at Solar Power
for Seattle." Proceedings of the Intersociety Energy
Conversion Engineering Conference (IECEC), pp. 360-
65, Aub. 1973.

1747 [No entry]

1748 "On Future Power from the Sun. " Chemistry 43: 25,
      March 1970. Illus.

1749 "On the Floor: Solar Energy. " Congressional Quart-
      erly 32(7): 381082, Feb. 16, 1974.

1750 "On the Floor: Solar Energy. " Congressional Quart-
      erly 32(21): 1403, May 25, 1974.

1751 "On the Floor: Solar Energy Research. " Congres-
      sional Quarterly 32(39): 2636-37, Sept. 28,
      1974.

1752 "On the Floor: Solar Heating, Cooling. " Congres-
      sional Quarterly 32(35): 2407-08, Aug. 31, 1974.

1753 "Optimism in Adapting the Sun's Energy for Mankind's
      Use. " Chemical & Engineering News 51(31): 14-17,
      July 30, 1973.

1754 Orlov, Vadim. "Hunting the Sun in the Steppes of
      Central Asia. " UNESCO Courier 27(1): 33-36, Jan.
      1974. Photos.

1755 Oswald, W. J. "Productivity of Algae in Sewage Dis-
      posal. " Solar Energy 15(1): 107ff (11p.), May 1973.
      Diagr., graphs, tables.

1756 Othmer, Donald F., and Oswald, A. Roels. "Power,
      Fresh Water, and Food from Cold, Deep Sea
      Water. " Science 183(4108): 121ff (5pp.), Oct. 11,
      1973.

1757 "Out of the Shadow. " Ecology Today 2(1): 29-31,
      April 1972.

1758 Page, J. K. "The Estimation of Monthly Mean Values
      of Daily Total Short-Wave Radiation on Vertical
      and Inclined Surfaces from Sunshine Records for
      Latitudes 40°N-40°S. " Proceedings of the United
      Nations Conference on New Sources of Energy 4:
      378+, 1964.

1758a Paige, Steven. Solar Water Heating for the Handyman.
      Barrington, NJ: Edmund Scientific, 1974. 31pp.
      Refs., illus. $4.00.

1759 Pakistan Solar Energy Commission. Solar Energy Prospects in Pakistan. Karachi: Golden Block Works. Illus.

1760 Patha, J. T., and Woodcock, G. R. "Feasibility of Large-scale Orbital Solar/Thermal Power Generation." Proceedings of the Intersociety Energy Conversion Engineering Conference (IECEC), pp. 312-319, 1973.

1761 Pawlik, E. V., et al. "Solar Electric Propulsion System Evaluation." Journal of Spacecraft and Rockets 7: 968-76, Aug. 1970. Refs., diagrs.

1762 Payne, Jack and McConaghie, Tom. "Solar Water Heater for Your Vacation Home." Popular Science 205(3): 104-105+, Sept. 1974. Diagrs., photos.

1763 Payne, P. A. Development and Pilot Line Production of Lithium Doped Silicon Solar Cells (Final Report). Sylmar, CA: Heliotek, 1972. 82pp. NTIS #N73-15078.

1764 Pearson, G. L. "Applications of Photovoltaic Cells in Communications." United Nations Conference on New Sources of Energy, E35-S40, Rome, 1961. New York: United Nations, 1964.

1765 _____. "Electricity from the Sun." Proceedings, World Symposium on Applied Solar Energy, Phoenix, Arizona, 1955. Menlo Park, CA: Stanford Research Institute, 1956.

1766 Peck, E. C. "Drying Lumber by Solar Energy." Sun at Work, 3rd Quarter, pp. 4-6, 1962.

1767 Pelletier, R. J. "Solar Energy: Present and Foreseeable Uses." Agriculture Engineering 40: 142-44, 151, 1959.

1768 Penrod, E. B., and Prasanna, K. V. "Procedure for Designing Solar-Earth Heat Pumps." Heating, Piping, and Air Conditioning 41(6): 97-100, June 1969. Diagrs., refs.

1769 Perrot, M. L'Energie Solaire. Paris: Fayard Bilandela Science, 1963.

1770 _____, Peri, G., and Robert, J. "A Thermoelectric Effect with Powdered Metallic Oxides." Transactions of the Conference on the Use of Solar Energy: The Scientific Basis 5: 36-42. Tucson: University of Arizona Press, 1958.

1771 _____, and Touchais, M. "Trends at IESUA for
Energy Production from Solar Radiation." United
Nations Conference on New Sources of Energy, E35-
S84, Rome, 1961.

1772 Petukhov, B. V. Solar Water Heating Installation.
Moscow, USSR: Academy Sciences, 1953. 42pp.

1773 Peyches, I. "Special Glasses and Mountings for the
Utilization of Solar Energy." United Nations Conference on New Sources of Energy, E S91, Rome,
1961. New York: United Nations, 1964-

1774 Peyturaux, Roger. L'Energie Solaire. Paris: Presses
Universitaires de France, 1968. 128pp. Illus.

1775 Pfeiffer, C., and Schoffer, P. "Performance of Photovoltaic Cells at High Radiation Levels." Transactions of the American Society of Mechanical Engineers, 85A, p. 208, 1963.

1776 _____, _____, Spars, B. G., and Duffie, J. A.
"Performance of Silicon Solar Cells at High Levels
of Solar Radiation." Transactions of the American
Society of Mechanical Engineers 84A: 33, 1962.

1777 "Photochemical Processes." Transactions of the Conference on the Use of Solar Energy: The Scientific
Basis 4 Tucson: University of Arizona Press, 1958.

1778 "Photovoltaics: Photons In/Electrons Out." Mosaic
5(2): 14-23, Spring 1974.

1779 Pickering, William H. "Important to Examine Possibilities of Unconventional Energy Sources." Aware
(31): 5-7, April 1973.

1780 Pillet, M. La Calculatrice Analogique du C. S. T. B. à
Champs sur Marne. Design Ltd.

1781 _____. "Les Zomes a l'Energie Solaire." L'Architecture d'Aujourd'hui No. 167: xxxix-xi, May 1973.

1782 Pirie, N. W. "The Use of Higher Plants for Storing
      Solar Energy. " Transactions, Conference on the
      Use of Solar Energy: The Scientific Basis 4: 115-
      21. Tucson: University of Arizona Press, 1958.

1783 Pivovonsky, M., and Nagel, M. R. Tables of Black-
      body Radiation Properties. New York: Macmillan,
      1961.

1784 Pleijel, G. V., and Lindström, B. I. "Stazione
      Astrofisica Svedese--A Swedish Solar-Heated House
      at Capri. " United Nations Conference on New
      Sources of Energy, E35-S49, Rome, 1961. New
      York: United Nations, 1964.

1785 "Plugging Into the Sun. " Science Digest 73(4): 31,
      April 1973. Drawing.

1786 Plumlee, R. H. "Perspectives in U. S. Energy Re-
      source Development. " Proceedings of the Interso-
      ciety Energy Conversion Engineering Conference
      (IECEC), Albuquerque, NM: Sandia Labs, 1973.
      113pp. NTIS #SLA-73-163/GA.

1787 Plunkett-Ernle-Erle-Drax, Sir Reginald Aylmer Ran-
      furly. A Handbook of Solar Heating, 3rd ed.,
      Poole (Dorset) J. Looker, 1965. 22pp. Illus.,
      tables, diagrs.

1788 Pope, R. B. and Schimmel, W. P., Jr. "An Analysis
      of Linear Focused Collectors for Solar Power. "
      Proceedings, 8th Intersociety Energy Conversion
      Engineering Conference (IECEC), Albuquerque, NM:
      Sandia Labs, 1972. 7pp. NTIS #SLA-73-5319/GA.

1789 _____, et al. "A Combination of Solar Energy and
      the Total Energy Concept: The Solar Community. "
      Proceedings of the 8th Intersociety Energy Conver-
      sion Engineering Conference (IECEC), Albuquerque,
      NM: Sandia Labs, 1973. 8pp. NTIS #SLA-73
      5318/GA.

1790 "Portable Solar Cooker Aimed at Campers. " Science
      Digest 75(4): 71, April 1974. Illus.

1791 Porteous, A. "Fresh Water for Aldabra." Engineering 209: 490, May 15, 1970.

1792 "Power From Sun: The Search Picks Up. " U.S. News
& World Report 74: 90-91, April 16, 1973. Photos.

1793 Prata, A. S. "A Cylindro-Parabolic Cooker." United
Nations Conference on New Sources of Energy, E35-
S11o, Rome, 1961. New York: United Nations, 1964.

1794 "The President's Materials Policy Commission Re-
sources for Freedom." Vol. 4. The Promise of
Technology. Washington, DC: U. S. Government
Printing Office, 1952.

1795 Prince, M. B. "Latest Developments in the Field of
Photovoltaic Conversion of Solar Energy." United
Nations Conference on New Sources of Energy,
E35-S65, Rome, 1961. New York: United Nations,
1964.

1796 Proceedings, Annual Power Conference, United States
Army Signal Research and Development Laboratory,
Red Bank, NJ: P. S. C. Publications Committee,
1960-64.

1797 Proceedings from the United Nations Conference on
New Sources of Energy, Vol. 4 (Solar Energy I,
$7. 50), Vol. 5 (Solar Energy II, $4. 50), Vol. 6
(Solar Energy III. $5. 00). From Sales Section,
United Nations, New York, 10017.

1798 Proceedings of Advanced Study Institute for Solar and
Aeolian Energy, Sounion, Greece. Greek Atomic
Energy Commission and the Hellenic Scientific Society of Solar and Aeolian Energy, 1961; In: Solar
and Aeolian Energy (Spanides, A. G. and Hatzikakidis, A. D. , eds.). New York: Plenum Press,
1964.

1799 Proceedings of Space Heating Symposium. Cambridge,
MA: Massachusetts Institute of Technology, 1950.

1800 Proceedings of the Solar Sea Power Plant Conference
and Workshop, Sponsored by the National Science
Foundation (RANN), Carnegie Mellon University,
Pittsburgh, PA: June 1973.

1801   "Proceedings of the Solar Furnace Symposium, Phoe-
       nix, Arizona." Solar Energy 1(2-3): 3-115, 1957.

1802   Proceedings of the World Symposium on Applied Solar
       Energy, Phoenix, Arizona, 1955. Menlo Park, CA:
       Stanford Research Institute, 1956.

1803   Procell, C. J. "Energy Conservation Systems: Follow-
       the-sun Concepts." Building Systems Design 69:
       15-16, May 1972.

1804   Proctor, D. "The Use of Waste Heat in a Solar Still,"
       Solar Energy 14: 433, 1973.

1805   "Project to Study Feasibility of 'Solar Climate Con-
       tro.'" Air Condition Heating and Refrigeration
       News 129: 34, June 11, 1973. Map.

1806   "Proposals Aim to Solve Energy Shortage." Chemical
       and Engineering News 50(15): 2, April 10, 1072.

1807   Putnam, P. C. Energy in the Future. Princeton:
       Van Nostrand, 1956.

1808   _____. "Expected World Energy Demands." Solar
       Energy Research (Daniels, F. and Duffie, J. A.,
       eds.). Madison: University of Wisconsin Press,
       1955, pp. 7-11.

1809   Pyramidal Optics: Energy Collection System. Stam-
       ford, CT: Wormser Scientific Corporation. 6pp.
       Photos., graphs.

1810   Quigg, P. W. "Alternative Energy Sources." Satur-
       day Review World 1(12): 31-32, Feb. 23, 1974.

1811   Raag, V. "Silicon-Germanium Thermocouple Develop-
       ment." 17th Annual Power Conference, Red Bank,
       NJ, U. S. Army Electronics, Research & Develop-
       ment Laboratory, 1963. pp. 34-36.

1812   Rabenhorst, D. W. "Superflywheel." Proceedings of
       the NSF/RANN Solar Heating and Cooling for Build-
       ings Workshop, NSF/RANN-73-004, pp. 60-68,
       July 1973.

1813   Rabinowitch, Eugene. "Challenges of the Scientific

Age. " Bulletin of Atomic Scientists 29(7): 4-8, Sept. 1973.

1814 _____. "Photochemical Redox Reactions and Photo-synthesis. " Photo-Chemistry in the Liquid and Solid States. (Heidt, L. J. , et al). New York: Wiley, 1960. p. 84.

1815 _____. "The Photochemical Storage of Energy. " Transactions of the Conference on the Use of En-ergy: The Scientific Basis 4: 182-87. Tucson: University of Arizona Press, 1958.

1816 _____. "Photochemical Utilization of Light Energy." National Academy of Science 47: 1296-1303; Solar Energy (Special Issue) 5: 52-58, Sept. 1961; Solar Energy Research (Daniels, F. and Duffie, J. A. , eds.); Madison: University of Wisconsin Press, 1955.

1817 _____. Photosynthesis and Related Processes. New York: Interscience, 1945 and 1951.

1818 Radin, Alex. "How We Can Conserve Energy. " Pub-lic Power 31(4): 6-9, July-Aug. 1973.

1819 Rahilly, W. Patrick. Theoretical Treatment of the Vertical Multijunctional Solar Cell. Wright-Patter-son AFB, OH: Air Force Aero Propulsion Lab, 1972. 42pp. NTIS #AD-754 901.

1820 Ralph, E. L. "A Commercial Solar Array Design. " Solar Energy 14(3): 279-286, Feb. 1973.

1821 _____. "A Plan to Utilize Solar Energy as an Electric Power Source. " IEEE Photovoltaic Spe-cialists Conference, pp. 326-30, 1970.

1822 _____. "Large Scale Solar Electric Power Gener-ation. " Solar Energy 14(1): 11ff (10pp.), Dec. 1972. Graphs, refs. , tables.

1823 Ramdas, L. A. "Solar Radiation and Its Measurement in India. " United Nations Conference on New Sources of Energy. , E35-S105, Rome, 1961. New York: United Nations, 1964.

1824 Ranney, M. Solar Cells. Park Ridge, NJ: Noyes
      Data Corp., 1969. $35.00. 271pp. Illus. $35.00.

1825 Rao, A. Bhaskara, and Padmanabhan, G. R. "Effect
      of Electrical Discharge on Silicon Solar Cell Char-
      acteristics." Applied Physics 3: 1849-54, Dec.
      1970. Refs., diagr.

1826 Rappaport, P. "Photoelectricity." Proceedings of the
      National Academy of Science 47: 1303-06, 1961.
      Also in Solar Energy (Special Issue 5: 59-62, Sept.
      1961.

1827 _____, and Moss, H. I. "Low Cost Photovoltaic
      Conversion of Solar Energy." United Nations Con-
      ference on New Sources of Energy, E35-S106,
      Rome, 1961.

1828 Rau, Hans. Solar Energy. Trans. by Maxim Shur,
      ed. and revised by J. A. Duffie. New York: Mac-
      millan, 1964. 171pp. Illus., portraits. $9.95.

1829 Ravich, L. E. "The Film Photovoltaic Devises for
      Solar Energy Conversion." United Nations Confer-
      ence on New Sources of Energy, E35-S56, Rome,
      1961. New York: United Nations, 1964.

1830 Ray, H. R. "Energy, Technology and Solararchitec-
      ture." Mechanical Engineering 95(11): 18ff (5pp.),
      Nov. 1973. Photos.

1831 "RCA in Manhattan Pioneers Project in Solar Heating."
      Interiors 133: 8, March 1974. Illus., plans.

1832 "RCA Scientist Predicts Widespread Home Use of Solar
      Energy in Five Years." Architectural Record 156
      (1): 35, July 1974. Photo.

1833 "Reference Energy Systems and Resource Data." As-
      sociated Universities, Inc., AET-8, April 1972.

1834 "Research Draws on Sun for Low-cost Air Condition-
      ing." Engineering News-Record 189: 14, Dec. 14,
      1972. Illus.

1835 "Research Frontiers in the Utilization of Solar Energy."
      Proceedings of National Academy of Science 47

(1961): 1245-1306. Also in Solar Energy (Special Issue) 5 Sept. 1961.

1836 "Research on Future Energy Systems. " Professional Engineering 42(2): 20ff (4-1/2pp.), Feb. 1972.

1837 Reynolds, D. C. "The Photovoltaic Effect in Cadmium Sulfide Crystals. " Transactions of the Conference on the Use of Solar Energy: The Scientific Basis 5: 102-16. Tucson: University of Arizona Press, 1958.

1838 Reynolds, John S. Solar Energy and Pacific Northwest Buildings. 59pp. $3.00.

1839 Rice, Randall G. Electrochemical Conversion Devices (A Report Bibliography). Arlington, VA: U. S. Defense Documentation Center, 1963. 140pp. Refs. 76-291866.

1840 Ridgeway, James. "Soaking Up the Sun. " Working Papers 1(4): 32ff. (5pp.), Winter 1974. Drawing.

1841 Rien, R. K. "Large Area Solar Cells Prepared on Silicon Sheet. " Proceedings of the 17th Annual Power Sources Conferences. Atlantic City, NJ: May 1963.

1842 Rink, J. E. , and Hewitt, J. G. , Jr. "Large Terrestrial Solar Arrays. " Proceedings of the Intersociety Energy Conversion Engineering Conference, Boston, MA, Aug. 3-5, 1971.

1843 Risser, Hubert E. "The U. S. Energy Dilemma: The Gap Between Today's Requirements and Tomorrow's. " Illinois State Geological Survey Report 64, July 1973. 59pp. Diagrs. , tables.

1844 Ritchie, D. W. , and Sandstrom, J. D. Multikilowatt Solar Laboratory. Pasadena, CA: Jet Propulsion Laboratory, California Institute of Technology, 1967. 14pp. Illus. (JPL Technical Report 32-1140.)

1845 Roberts, Ralph. "Energy Sources and Conversion Techniques. " American Scientist 61(1): 66ff (10pp.), Jan. -Feb. 1973. Diagrs. , graphs, photo. , refs. , tables.

1846 Robinson, N. A Brief History of Utilization of the
Sun's Radiation. Jerusalem: Actes de Septieme
Congres, International d'Histoire des Sciences, 1953.

1847 _____. Proceedings of World Symposium on Applied Solar Energy, Phoenix, Arizona, 1955. Menlo
Park, CA: Stanford Research Institute, 1956. pp.
43-61.

1848 _____. A Report on the Design of Solar Energy
Machines, NS/AZ/141. Paris: UNESCO, 1953.

1849 _____. Solar Radiation. New York and Amster-
dam: Elsevier, 1962.

1850 _____, and Neeman, E. "The Solar Switch, An
Automatic Device for Economizing Auxiliary Heating of Solar Water Heaters." United Nations Conference on New Sources of Energy, E35-S31, Rome,
1961. New York: United Nations, 1964.

1851 Rodale, R. "Why Organic Growers Love the Sun."
Organic Gardening and Farming 18: 30-35, Oct.
1971. Illus.

1852 Roddis, Louis H., Jr. "Unconventional Power Sourc-
es." Survey Report presented at Edison Electric
Institute Executives Conference, Scottsdale, AZ:
Dec. 4, 1970.

1853 Rogers, Benjamin T. "Using Nature to Heat and
Cool." Building Systems Design 70(7): 11ff (5pp.),
Oct.-Nov. 1973. Diagrs., graphs, photos., tables.

1854 Rollefson, G. K., and Burton, M. Photochemistry.
Englewood Cliffs, NJ: Prentice-Hall, 1939.

1855 "Roll-up Solar Array Systems Developed." Aviation
Week & Space Technology 92: 95, June 22, 1970.
Illus.

1856 Root, D. E., Jr. "Practical Aspects of Solar Swimming Pool Heating." Solar Energy 4(1): 23, 1960.

1857 Roscow, J. P. "Sun Belongs to Everyone, But Putting
It to Work Is Expensive." Financial World 140:
14-15+, Aug. 1, 1973. Illus.

1858 Rossini, F. D., et al. "Selected Values of Chemical Thermodynamic Properties." National Bureau of Standards Circular 500, 1952. Repr. 1961.

1859 Ruchlis, Hyman. Thank You, Mr. Sun! Irvington-on-Hudson, NY: Harvey House, Inc., 1957. 43pp. Illus. PLB $3.36, $3.50. (Gr. 2-5).

1860 Russell, John L., Jr. "Investigation of a Central Station Solar Power Plant." Proceedings of the Solar Thermal Conversion Workshop, Washington, DC: Jan. 1973. Also published as Gulf General Atomic Report No. Gulf-GA-A12759, Aug. 31, 1973.

1861 _____, De Plomb, E. P., and Ravinder, R. K. "Principles of the Fixed Mirror Solar Concentrator." Solar Energy, submitted for publication.

1862 Saar, R. A. "Solar Energy Collector Accepts Diffused Sunlight." Physics Today 27(8): 19-20, Aug. 1974.

1863 Sahai, R., and Milnes, A. G. "Heterojunction Solar Cell Calculations." Solid State Electronics 13: 1289-99, Sept. 1970. Refs., diagrs.

1864 Sakurai, T., Osamu, K., Koro, S., and Koji, I. "Construction of a Large Solar Furnace." Solar Energy 8(4): 117-26, Oct. 1964.

1865 Salam, E., and Daniels, F. "Revêtement Rayonnant de Façon Selective pour le Chauffage Solaire." Applications Thermiques de l'Energie Solaire dans le Domaine de la Recherche et de l'Industrie, Mont-Louis, France, 1958. Paris: Centre National de la Recherche Scientifique, 1961. pp. 483-93.

1866 _____, and _____. "Solar Distillation of Salt Water in Plastic Tubes Using a Flowing Air Stream." Solar Energy 3(1): 19-22, Jan. 1959.

1867 "Salin Water Conversion." Advances in Chemistry, Series 27. Washington, DC: American Chemical Society, 1960.

1868 Salisbury, David F. "Tapping the Sun's Energy." Christian Science Monitor p. 5, Nov. 29, 1972.

1869 Sancier, K. M. "Photo-Galvanic Cells." Transactions of the Conference on the Use of Solar Energy: The Scientific Basis 5: 43-56.

1870 Sandstrom, Jerome D. A Method for Predicting Solar and Cell Current-Voltage Curve Characteristics as a Function of Incident Solar Intensity and All Temperature. Pasadena, CA: Jet Propulsion Laboratory, California Institute of Technology, 1967. 7pp. (JPI 32-1142) 68-1764.

1871 Sapsford, C. M. "Solar Radiation Records in Australia and Their Presentation." United Nations Conference on New Sources of Energy, E35-S32, Rome, 1961. New York: United Nations, 1964.

1872 Sargent, S. L. "A Compact Table of Blackbody Radiation Functions." Bulletin of the American Meteorological Society 53: 360+, April 1972.

1873 _____, and Beckman, W. A. "Theoretical Performance of an Ammonia- Sodium Thiocyanate Intermittent Absorption Refrigeration Cycle." Solar Energy 12(1): 137-146, Jan. 1968.

1874 Sarles, F. W., Jr., et al. "Balloon Flight Instrumentation for Solar Cell Measurements." Review of Scientific Instruments 42: 346-53, March 1971. Diagrs.

1875 Satcunanthan, S., and Deonarine, S. "A Two-Pass Solar Air Heater." Solar Energy 15(1): 41-49, 1973.

1876 "Satellite Follows the Sun." Saturday Review of the Sciences 1(2): 59, March 1973. Diagr.

1877 "Satellites Make Electricity from Sunlight." Science Digest 75(4): 76-71, April 1974. Photo., drawing.

1878 Savornin, J. "Efficiency of Various Types of Solar Stills." Transactions of the Conference on the Use of Solar Energy: The Scientific Basis 3: 134-37. Tucson: University of Arizona Press, 1958.

1879 _____. "Study of Solar Water Heating in Algiers." United Nations Conference on New Sources of Energy, E35-S72, Rome, 1961. New York: United Nations, 1964.

1880 Schaeper, H. R. A., and Farber, Erich A. "The
Solar Era: Part IV--The University of Florida
Electric." Mechanical Engineering 94(11): 20-24,
Nov. 1972. Photos., diagrs., refs., graphs.

1881 Schimmel, W. P., Jr. "A Vector Analysis of the
Solar Energy Reflection Process." Annual Meeting
of the U. S. Section of the International Solar Energy Society, Cleveland, OH, Oct. 1973.

1882 Schmidt, R. N., Park, K. C., and Janssen, E.
"High Temperature Solar Absorber Coatings, Part
II." Tech. Doc. Report No. ML-TDR-64-250 from
Honeywell Research Center to Air Force Materials
Laboratory, Sept. 1964.

1883 Schoffer, P., Kuhn, P., and Sapsford, C. M. "Instrumentation for Solar Radiation Measurement."
United Nations Conference on New Sources of Energy, E35-S92, Rome, 1961. New York: United
Nations, 1964.

1884   , and Pfeiffer, C. "Performance of Photovoltaic Cells at High Radiation Levels." American
Society of Mechanical Engineers, Paper 62-WA-
348, 1963.

1885   , and Suomi, V. E. "A Direct Current Inte-
grator for Radiation Measurements." Solar Energy
5(1): 29, 1961.

1886 Scholes, W. A. "Solar Stills for Sheep: Solar-Powered
Desalination Equipment." Science News 92(9): 204₊,
Aug. 26, 1967.

1887 Schrenk, G. L., and Gritton, D. C. Final Report,
Analysis of Solar Reflectors. Mathematical Theory
and Methodology for Simulation of Real Reflectors.
Indianapolis, IN: Allison Division, General Motors
Corp., 3693 Contract AF04 (6950335), Dec. 1963.

1888 Science 184(4134): Entire Issue devoted to Energy,
April 19, 1974.

1889 "Scientists Call for Solar Energy in Housing." Engineering News-Record 191: 14, July 19, 1973.

1890 "Scientists Say Electricity from Sun (Feasible)."

Energy Digest 2(5): 5ff (4pp.), Sept. -Oct. 1973, 1973. Diagrs., graph.

1891 Scott, D. "World's Biggest Furnace Runs on Sunshine." Popular Science 196[2]:88-9, Feb. 1970. Illus.

1892 Scott-Monck, J. A., et al. Design and Fabriaction of Wraparound Contact Silicon Solar Cells. Sylmar, CA: Heliotek, 1972. 30pp. NTIS #3N73-13051.

1893 Seaborg, Glenn T., et al. "Nuclear and Other Energy." NY Academy of Sciences Annals 216: 79ff (10pp.), May 18, 1973. Table.

1894 Seibert, D. "Building: First House to Use Sun for Power." House Beautiful 115: 18+, Oct. 1973. Illus.

1895 Sekihara, K. "Solar Radiation in Japan." United Nations Conference on New Sources of Energy, E35-S2, Rome, 1961. New York: United Nations, 1964.

1896 Selcuk, M. K. "Flat Plate Solar Collector Performance at High Temperatures." Solar Energy 8(2): 57-62. April, 1964.

1897 _____, and Yellot, J. K. "Measurement of Direct, Diffuse, and Total Radiation with Silicon Photovoltaic Cells." Solar Energy 6(4): 155-63, Oct. 1962.

1898 _____, and Ward, G. T. "Optimization of Solar Terrestrial Power Production Using Heat Engines." Journal of Engineering for Power 92: 173-81, April 1970. Refs., charts, diagrs.

1899 Seraphin, B. O. "Heat Transfer Characteristics of a Linear Solar Collector." Applied Optics 12: 349-54, Feb. 1973. Diagrs.

1900 Seybold, R. "Fusible Salts and Nitrogen Dioxide Adsorption for Utilizing Solar Energy." B.S. Thesis, University of Wisconsin, 1956.

1901 Shaffer, L. H. "Wavelength Dependent (Selective) Processes for the Utilization of Solar Energy." Solar Energy 2(3-4): 21-26, July, 1958.

1902 Shapley, D. "House, Senate Differ on Solar Energy."
Science 184(4132): 49, April 5, 1974.

1903 Shaw, G. A. D., and Fukoner, J. D. "SERT II: Solar
Array Power System." Journal of Spacecraft and
Rockets 8: 231-33, March 1971. Illus., diagrs.

1904 Shefter, Ya., et al. "Solar and Wind Power to be
Harnessed (Energiyu Solntsa I Vetra-V
UPRYAZHU)." Pravda, Nov. 1, 1972. 5pp. Trans
by Army Foreign Science and Technology Center,
Charlottesville, VA. NTIS #AD-765 783/6.

1904a "Shell Platforms Utilize Wind and Sun Power." Petroleum Engineer International 46: 14, Sept. 1974.

1905 Shepard, N. F., et al. Feasibility Study of a 100
Watt per Kilogram Lightweight Solar Array System
(Final Report). Philadelphia: General Electric,
1973. 315pp. NTIS #N73-29006/6GA.

1906 _____, Stahle, C., Schneider, A., and Hanson,
K. L. Feasibility Study of a 110 Watt Per Kilo-
gram Lightweight Solar Array System (Quarterly
Report 1 Aug. -31 Oct. 1972). Philadelphia: General
Electric. 110pp. NTIS #N73-15079.

1907 Sheridan, Norman R. "Criteria for Justification of
Solar Energy Systems." Solar Energy 13(4): 425ff
(4-1/2pp.), July, 1972.

1908 _____. "Performance of the Brisbane Solar House."
Paper presented at International Solar Energy Society Conference, Melbourne, 1970.

1909 _____. "Prospects for Solar Air Conditioning in
Australia." United Nations Conference on New
Sources of Energy, E35-S39, Rome, 1961. New
York: United States, 1964.

1910 _____, Bullock, K. M., and Duffie, J. A. "Study
of Solar Processes by Analog Computer." Solar
Energy 11(2):69-77, April 1967.

1911 Shoemaker, M. J. "Notes on a Solar Collector with
Unique Air Permeable Media." Solar Energy 5(4):
138-41, Oct. 1961.

1912   Shriner, Robert D. and Cohen, Michael. "A Solar
        Energy Bibliography." Professional Engineer 43
        (10): 17-19, Oct. 1973.

1913   Shumann, W. A. "Man Turns to Sun as Energy Alter-
        native." Aviation Week & Space Technology 100:
        56-57+, Jan. 14, 1974.

1914   Shurcliff, W. A. Solar Heated Buildings: A Brief
        Survey. 6th ed. Published by Solar Energy Digest,
        Oct. 17, 1974. (Available from Solar Energy Di-
        gest, 19 Appleton St., Cambridge, MA 02138.

1915   "Sieze That Sunbeam." National Observer p. 1, July
        8, 1972.

1916   Silk, Leonard. "Nation's Energy Crisis." New York
        Times, p. 57, April 26, 1972.

1917   Simon, A. W. "Loss of Energy by Absorption & Re-
        flection in the Heliostat and Parabolic Condenser of
        a Solar Furnace." Solar Energy 2(2): 30-33, 1958.

1918   "Sixth in a Series, Solar Alternative to Atomic Power."
        Science Digest 39(3): 31ff (6pp.), March 1972.

1919   "Skylab Power Augmentor in Development." Aviation
        Week & Space Technology 99: 49-50, July 2, 1973.

1920   Smernoff, Barry J. "Energy Policy Interactions in the
        United States." Energy Policy 1(2): 136ff (18pp.),
        Sept. 1973. Diagr., graph, refs., table.

1921   Smil, Vaclav. "Energy and the Environment." Fu-
        turist 8(1): 4-13, Feb. 1974. Photos., diagrs.

1922   Smith, Gene. "Energy From Sun Seen in 5 Years."
        New York Times, June 5, 1974.

1923   Smoleniec, S. Utilization of Solar Energy for Air Con-
        ditioning. Johannesburg: Witwatersrand University
        Press, 1962. 29pp. (72-18560)

1924   Snyder, Nathan W., ed. Space Power Systems: A
        Selection of the Technical Papers Based Mainly on
        a Symposium of the American Rocket Society, Santa
        Monica, CA, 1966. New York: Academic Press,

1966. 632pp. Illus., diagrs. Also in Progress in Astronautics and Rocketry, Vol. 4. 61065038

1925 Snyder, R. E. "Solar Power Generator Cuts Offshore Operating Cost." World Oil 176: 81-83, May 1973. Illus.

1926 Sobotka, R. "Economic Aspects of Commercially Produced Solar Water Heaters." Solar Energy 10(1): 9-14, Jan. 1966.

1927 _____. "Solar Water Heaters." United Nations Conference on New Sources of Energy, E35-S26, Rome, 1961. New York: United Nations, 1964.

1927a "Sohio Petroleum Puts Another Foot on the New Solar Energy Bandwagon." Chemical Marketing Reporter 206: 7+, Oct. 7, 1974.

1928 "'Sola,' Sfera luminosa ad accensione solare." Domus (509): 23b April 1972. Illus., diagrs.

1929 "Solar Array Features Low Volume and Mass." Electronic Engineering 42: 24-25, July 1970. Illus.

1930 "Solar Arrays' Particle Resistance Tested." Aviation World 94: 66-67, June 14, 1971.

1931 "Solar Battery Charger Aids Line Research." Electrical World 172: 49, Nov. 17, 1969. Illus.

1932 "Solar-cell Efficiency Boosted to 18 Per Cent." Machine Design 44: 10, June 1, 1972.

1933 "Solar Cell Helps Make Patient Monitoring Safe." Electronics 44: 22-23, May 24, 1971. Diagr.

1934 Solar Cells. Proceedings of the International Colloquium Organized by the European Cooperation Space Environment Committee July 6-10, 1970, Toulouse, France with the collaboration of the Centre National d'Etudes Spatieles (CNES) and the Department d'Etudes et de Recherches, Toulouse. London, New York: Gordon and Breach Science Publishers, 1971. 680pp. Illus., refs.

1935 Solar Cells and Solar Panels 1958-73. (Bibliography).

Alexandria, VA: Defense Documentation Center, 1973. 230pp. NTIS #AD-768 400/4GA.

1936 "Solar Cells for Home Appliances Are Reality." Product Engineering 40: 16, Nov. 17, 1969.

1937 "Solar Cells Power Light Navaids." Electronics & Power 19 370, Sept. 6, 1973. Illus.

1938 "Solar Climate Control--New Industry." Progressive Architecture 54: 44, July 1973.

1939 "Solar Climate Control Markets Under Study." Industry Week 177: 21+, May 21, 1973.

1940 Solar Cooker Construction Manual, VITA, June 1967.

1941 Solar Effects on Building Design. Washington, DC: Building Research Institute, 1963.

1942 "Solar Energy." Architectural Design 42: 777, Dec. 1972. Illus.

1943 Solar Energy (Vol. 1). Phoenix, AZ: Association for Applied Solar Energy, Jan. 1957.

1944 "Solar Energy." Encyclopedia of Science and Technology, 12. New York: McGraw-Hill, 1960, p. 467.

1945 Solar Energy. Rockville, MD: Intern'l Solar Energy Soc., Smithsonian Radiation Biology Lab. Back issues: Elmsford, NY: Pergamon Press.

1945a "Solar Energy Applications." ASHRAE Journal 16(9): 29-49, Sept. 1974. Refs., flow diagr., illus., map, diagrs.

1946 "Solar Energy as a National Energy Resource." NSF/NASA Solar Energy Panel, Dec. 1972 submission to the White House Office of Science and Technology. For copy, write William Cherry, Solar Research Division, NASA, Goddard Space Center, Greenbelt, MD 20771.

1947 "Solar Energy Bills Stalled in Congress While Some Collectors Go On Line." Engineering News-Record 192: 10, March 28, 1974.

1948　"Solar Energy: Boon to Developing Nations?" Machine
　　　　Design 44: 36, May 18, 1972.

1949　"Solar Energy Conference. " Congressional Quarterly
　　　　32(24): 1567, June 15, 1974.

1950　Solar Energy Evaluation Group Report. Argonne Na-
　　　　tional Lab, IL 1973. 50pp. NTIS #ANL-8045/GA.

1950a　"Solar Energy Experiment Will Get Commercial Ef-
　　　　fort. " Chemical Marketing Reporter 206(10): 7+,
　　　　Sept. 2, 1974.

1951　"Solar Energy: Heating and Cooling of Buildings: Sol's
　　　　Comfort Package. " Mosaic 5(2): 14-23, Spring
　　　　1974.

1952　"Solar Energy in Developing Countries: Perspective
　　　　and Prospects. " A report of an ad hoc advisory
　　　　panel to the National Academy of Science, Wash-
　　　　ington (March 1972).

1952a　"Solar Energy May Achieve Wide Use by 1980's. "
　　　　Chemical & Engineering News 51(5): 12-14, Jan.
　　　　29, 1973. Diagr. , photo.

1953　"Solar Energy: Practical Power Source for Earth?"
　　　　Machine Design 44: 6, Aug. 1972. Illus.

1953a　"Solar Energy Programme Moves into Second Phase. "
　　　　Engineer 239: 16, Sept. 5, 1974.

1954　"Solar Energy--Prospects for Its Large-Scale Use. "
　　　　Science Teacher 39(3): 36ff (4pp.), March 1972.

1955　"Solar Energy R & D Edges Ahead. " Public Utilities
　　　　Fortnightly 90: 48-49, Aug. 17, 1972.

1956　"Solar Energy Research--A Multi-Disciplinary Ap-
　　　　proach. " Staff Report of the Committee on Science
　　　　and Astronautics of the House of Representatives,
　　　　Dec. 1972.

1957　"Solar Energy Researchers Try a New Way. " Business
　　　　Week (2180): 72, June 12, 1971. Photo.

1958　"Solar Energy Science Building Will Open This Year. "

Architectural Record 156(1): 37, July 1974. Photo.

1959 "Solar Energy: Stoking the Boilers with Sunshine."
Mosaic 5(2): 14-23, Spring 1974.

1960 Solar Energy Studies. Recent Reports on Research in
the Field of Solar Energy in Progress at the University of Florida, Gainesville, FL. Engineering
and Industrial Experiment Station, College of Engineering, University of Florida, 1960. 28pp. Illus.,
portraits, maps, diagrs. Technical Report #9.
A60-9485.

1961 "Solar Energy Studies Requested for Heating, Cooling
Buildings." Air Conditioning Heating & Refrigeration News 129(22): 4, May 28, 1973.

1961a "Solar Energy System to Handle School's Heating and
Cooling Loads." Machine Design 46(20): 8, Aug.
22, 1974.

1962 "Solar Flat Plate Collector and Tank." Architectural
Design 42(7): 426, July 1972.

1963 "Solar, Geothermal Energy Agencies: Are They Needed?" Congressional Quarterly 32(20): 1227-78,
May 1, 1974.

1963a "Solar Heat from Patio Panels." Factory 7: 21,
Nov. 1974.

1964 "Solar Heater." Mother Earth News (21): 57, May
1973. Photo., diagr.

1965 Solar Heating and Cooling Demonstration Act. 93rd
Congress 1st Session HR 10952, Nov. 1973. 502pp.

1966 "Solar Heating, Cooling Planned for Office Building."
Air Conditioning, Heating & Refrigeration News 130:
26, Sept. 10, 1973.

1967 "Solar Heating for the Handyman." Mechanics Illustrated 70(556): 15-19, Sept. 1974. Diagrs.

1968 "Solar Is Safer, Gunter Contends." Congressional
Quarterly 32(29): 1879, July 20, 1974.

1969 "Solar Lab Uses Space-Derived Expertise. " Aviation
Week & Space Technology 100: 56-57, Feb. 4, 1974.

1970 "Solar Power Cells, How They Work in Space. " Space
World J-10(118): 25, Oct. 1973.

1971 "Solar-Power Conference, Paris: Abstracts of Papers."
Electrical World 180: 24, Aug. 1, 1973.

1972 "Solar Power Has Bright Future. " Industry Week 179:
35-37, Oct. 29, 1973. Map, illus.

1973 The Solar Power Option. Meeting of American Physics
Teachers, San Francisco, Feb. 2, 1972. 15pp.

1974 "Solar Power Outlook Heats Up. " Industrial Research
15: 24, March 1973.

1975 "Solar Power Satellites Could Ease Energy Crisis. "
Chemical & Engineering News 51(1): 17, Jan. 1,
1973. Table.

1976 "Solar Power Station. " Architectural Design 43(42):
142, March 1972. Illus., diagrs.

1977 "Solar-Powered Engineering Building. " Chemistry 47:
23, Feb. 1974. Illus.

1978 "Solar Powered History. " High Country News 6(7):
8-9, March 29, 1974. Photos., drawings.

1979 "Solar Powered Houses Arrive at Last. " Science Digest 74(3): 60-62, Sept. 1973. Diagr., drawing.

1979a "Solar Powerplants Would Be Great But.... " Engineering News-Record 193(21): 42, Nov. 14, 1974.

1980 "Solar R & D Is Bright. " Engineering News-Record
193(14): 13, Sept. 26, 1974.

1980a "Solar Radiation, Fusion: Best for Energy Needs. "
Industrial Research 14: 32, Aug. 1972. Illus.

1981 Solar Water Heaters, Division of Mechanical Engineer-
ing Circular No. 2. Available from Div. of
Mechanical Engineering, P.O. Box 26, Graham
Rd. , Highett, Victoria, Australia 3190.

1982 "Some Aspects of R and D Strategy in the United States." Electrical Review 193(7): 222-26, Aug. 17, 1973. Diagr., photos.

1983 Spanides, A. G., and Hatzikakidis, Nathan D., eds. International Seminar on Solar and Aeolian Energy, Sounion, Greece, 1961. Athens: NATO Institute of Advanced Studies in Solar and Aeolian Energy. Distr. by Plenum Press, NY, 1964. 491pp. Illus., diagrs., tables. $25.00.

1984 Speyer, E. "Optimum Storage of Heat with a Solar House." Solar Energy 3(4): 24, Dec. 1959.

1985 _____. "Solar Buildings in Temperate and Tropical Climates." United Nations Conference on New Sources of Energy, E35-S8, Rome, 1961. New York: United Nations, 1964.

1986 Spiegler, K. S. "Solar Evaporation of Salt Brines in Open Pans." Solar Energy Research (Daniels, F. and Duffie, J. A., eds.) Madison: University of Wisconsin Press, 1955.

1987 Spies, Henry R., et al. 350 Ways to Save Energy (and Money) in Your Home and Car. New York: Crown, 1974. 143pp.

1988 Sporn, P., and Ambrose, E. R. "The Heat Pump and Solar Energy." Proceedings, World Symposium on Applied Solar Energy, Phoenix, Arizona, 1955. Menlo Park, CA: Stanford Research Institute, 1956. pp. 159-70.

1989 Srivastava, Ashok. "Effect of Electrical Discharge on the Behavior of Silicon Solar Cells." IEE-IERE Proceedings (India) 10(4): 122-26, July-Aug. 1972.

1990 Stam, H. "Cheap But Practical Solar Kitchens." United Nations Conference on New Sources of Energy, E35-S24, Rome, 1961.

1991 Stanley, A. G. "Comparison of Low-energy Proton Damage in Ion-implanted and Diffused Silicon Solar Cells." IEEE Proceedings 59: 321-22, Feb. 1971. Refs.

1992   Starr, Chauncey. "Research on Future Energy Sys-
       tems." Professional Engineer 42(2): 20-24, Feb.
       1972. Chart, drawings.

1993   Stehling, Kurt R. "Turn On the Sun." NOAA 492:
       4ff (6pp.), April 1974. Drawings, photos.

1994   Steward, W. Gene. "A Concentrating Solar Energy
       System Employing a Stationary Spherical Mirror
       and Movable Collector." Proceedings of the NSF/
       RANN Solar Heating and Cooling for Buildings
       Workshop, NSF/RANN-73-004, pp. 17-20, July
       1973.

1995   Stoner, Carol H. "Sun: A Potential Powerhouse: Elec-
       tricity from the Sun." Organic Gardening and
       Farming 21(2): 24-25+, Feb. 1974.

1996   _____. "Sun: A Potential Powerhouse: Methane Gas
       from Algae." Organic Gardening and Farming 21:
       104+, April 1974.

1997   _____, ed. Producing Your Own Power: How To
       Make Nature's Energy Sources Work for You. Em-
       maus, PA: Rodale Press, Sept. 1974. 229pp.
       Illus., refs., index.

1998   Stonier, Tom. "An International Solar Energy Develop-
       ment Decade: A Proposal for Global Cooperation."
       Bulletin of the Atomic Scientists 18(5): 31-35, May
       1972.

1999   Streed, Elmer R. "Some Design Considerations for
       Flat Plate Collectors." Proceedings of the NSF/
       RANN Solar Heating and Cooling for Buildings Work-
       shop, NSF/RANN-73-004, pp. 26-35, July 1973.

2000   Strobel, J. J. "Developments in Solar Distillation--
       United States Department of Interior." United Na-
       tions Conference on New Sources of Energy, E35-
       S85, Rome, 1961. New York: United Nations, 1964.

2001   Stromberg, R. P. Solar Community, Albuquerque,
       NM, Sandia Labs (N.D.) 8pp. NTIS #SLA-73-5411/
       GA.

2002   Strother, John A. "Solar Energy Conversion." IEEE

Spectrum rp. 11-12, April 1971.

2003   Struss, R. G.   "Solar Furnace Determination of High
        Temperature Absorption Co-efficients. " Solar En-
        ergy 4(2): 21-26, April 1960.

2003a   "[Sulfur trioxide] SO₃ Dissociation Keys Solar Energy
        System. " Chemical and Engineering News 52(31):
        21, Aug. 5, 1974. Diagr.

2004   Sullivan, M. "World's Largest Solar Furnace Focuses
        Heat for Pure Materials. " Product Engineering 41:
        13-14, Jan. 19, 1970. Illus.

2005   Sumner, J. "Batteries Bathe As the Solar Cell Be-
        comes a Reality. " Engineer 237: 57, Nov. 15,
        1973. Diagr.

2005a   _____ . "Let the Sun's Energy Shine on Water Heat-
        ing. " Engineer 239: 17,   Sept. 26, 1974.

2006   "Sun Aids Russia's Materials Testing. " Industrial Re-
        search 16: 31, Dec. 1974.

2007   Sun at Work. Tempe, AZ: Solar Energy Society (form-
        erly the Association for Applied Solar Engery.)
        Quarterly magazine.

2008   "Sun City. " Architectural Forum 139: 72, Dec. 1973.

2009   "Sun Could Brighten Long-Term Energy Picture for
        U. S. Economy. " Commerce Today 4: 4-7, March
        18, 1974. Illus.

2010   "Sun Discovered: Diverse U. S. Agencies Crank Up
        Special Solar-energy Studies. " Engineering News-
        Record 192(5):11, Jan. 31, 1974.

2011   "Sun House, Norvège. " L'Architecture d'Aujourd'hui
        (163): 94-95, Aug. 1972. Illus., diagrs.

2012   "Sun in the Service of Man. " American Academy of
        Arts and Sciences 79: 181-326, 1951.

2013   "Sun May Provide Power in Developing Countries. "
        IEEE Spectrum 9: 86-87, May 1972.

2013a "Sun Power: A Far-Out Idea Comes Down to Earth. "
Chemical Week 115(25): 38-39, Dec. 18, 1974.
Photo.

2014 "Sun Power in the Pyrenees. " Time 95: 52-55, May
18, 1970. Illus.

2015 "Sun Power to be Beamed to Earth. " Machine Design
44: May 4, 1972.

2016 "Sun Will Heat and Cool Office Building. " Engineering
News-Record 191: 18, Aug. 16, 1973. Illus.

2017 "Sunlight and Bodies Heat This School. " Science Digest 69: 16-17, Jan. 1971. Illus.

2018 "Sun-Powered Furnace. " Chemistry 43: 23, Nov. 1970.
Illus.

2019 "Sunshine on Plants: Zero Discharge Goal. " Chemical
Week 112(20): 57-59, May 16, 1973. Photos.

2020 "Sunshine Trap Could Produce Nation's Power. "
Machine Design 43: 12, June 10, 1971.

2021 Suomi, V. E. , and Kuhn, P. M. "An Economical Net
Radiometer. " Tellus 10(1): 168, 1958. Also in
Quarterly Journal, Royal Meterological Society 84:
134, 1958.

2022 Swartman, R. K. "Solar Power--The Clean Energy. "
Engineering Journal 56(1): 14-18, Jan. 1973. Diagr.,
photo. , refs.

2023 _____, and Swaminathan, C. "Further Studies
on Solar-Powered Intermittant Absorption Refrigeration. " Paper presented at International
Solar Energy Society Conference, Melbourne, 1970.

2024 _____, and _____. "Solar Powered Refrigeration. " Mechanical Engineering 93: 22-24, June 1971.

2025 _____, et al. "The Solar Era Part 5: The Pollution of Our Solar Energy. " Mechanical Engineering
94(12): 23-27, Dec. 1972. Drawings.

2025a Swet, C. J. "Prototype Solar Kitchen. " Mechanical

Engineering 96(8): 32-35, Aug. 1974. Refs., diagrs.

2026 _____. A Universal Solar Kitchen. Silver Springs, MD: Johns Hopkins University Applied Physics Laboratory, 1972. 27pp. NTIS #PB-213 023/5.

2027 Swinbank, W. C. "Long-Wave Radiation from Clear Skies." Quarterly Journal of the Royal Meteorological Society 89: 1963.

2028 Szczelkun, Stefan A. "Do-It-Yourself-And-Save-Yourself-Sun-and-Wind Energy Handbook." Intellectual Digest 4(10): 23-30, June 1974. Drawings, plans, charts.

2029 Szego, C. G. "The U. S. Energy Problem, Vol. II: Appendices--Part A." NTIS Report PB-207 518, Nov. 1973. 744pp. Diagrs., graphs, tables.

2030 _____, Fox, J. A., and Eaton, D. R. "The Energy Plantation." Paper no. 729168. Proceedings of the Intersociety Energy Conversion Engineering Conference (IECEC), p. 113-14, Sept. 1972.

2031 Szokolay, S. and Hobbs, R. "Using Solar Energy in Housing." RIBA Journal 80: 177-79, April 1973. Refs., diagr.

2032 Tabor, H. "Large-Area Solar Collectors (Solar Ponds) for Power Production." United Nations Conference on New Sources of Energy, E35-S47, Rome, 1961. New York: United Nations, 1964. Also in Solar Energy 7: 189-94, 1963.

2033 _____. "A Note on the Thermo-syphon Solar Hot Water Heater." Bulletin, Cooperation Mediterranéenne pour l'Energie Solaire (COMPLES), No. 17, 33, 1969.

2034 _____. "Selective Radiation I. Wavelength Discrimination." Transactions of the Conference on the Use of Solar Energy: The Scientific Basis 2 (1A): 24-33. Tucson: University of Arizona Press, 1958.

2035 _____. "Selective Radiation II. Wave-Front Discrimination." Transactions of the Conference on the Use of Solar Energy: The Scientific Basis 2(1A): 34-40. Tucson: University of Arizona Press, 1958.

2036 _____. "Solar Collectors, Selective Surfaces, and Heat Engines." Proceedings of the National Academy of Science 47: 1271-78, 1961. Also in Solar Energy (Special Issue) 5 Sept. 1961.

2037 _____. "Solar Energy Collector Design." Transactions of the Conference on Use of Solar Energy: The Scientific Basis 2. Tucson: University of Arizona Press, 1958.

2038 _____. "Solar Energy Research: Program in the New Desert Research Institute in Beersheba." Solar Energy 2(2):3-6, April 1958.

2039 _____. "Stationary Mirror Systems for Solar Collectors." Solar Energy 2(3-4):27-33, July 1958.

2040 _____. "Use of Solar Energy for Mechanical Power and Electricity Production by Means of Piston Engines & Turbines." United Nations Conference on New Sources of Energy, E35-Gr-59, Rome, 1961. New York: United Nations, 1964.

2041 _____, and Bronicki, J. L. "Small Turbine for Solar Energy Power Package." United Nations Conference on New Sources of Energy, E35-S54, Rome, 1961. New York: United Nations, 1964.

2042 _____, Harris, J., Weinberger, H., and Doron, B. "Further Studies on Selective Black Coatings." United Nations Conference on New Sources of Energy, E35-S46, Rome, 1961. New York: United Nations, 1964.

2043 _____, and Matz, R. "Solar Pond Project." Solar Energy 9(4):177-182, Oct. 1965.

2044 _____, and Zeimer, H. "Low-Cost Focusing Collector for Solar Power Units." United Nations Conference on New Sources of Energy, Special preprint, Rome, 1961. New York: United Nations, 1964.

2045 Tabor, J. C. "Use of Solar Energy for Cooling Purposes. " United Nations Conference on New Sources of Energy, E35-Gr-S18, Rome, 1961. New York: United Nations, 1964.

2046 Talbert, S. G. , Eibling, J. A. , and Löf, G. O. G. Manual on Solar Distillation of Saline Water. Washington: Office of Saline Water, U. S. Dept. of Interior, Research and Development Progress Report #546, 1970.

2047 Talwalker, A. T. , Duffie, J. A. , and Löf, G. O. G. "Solar Drying of Oil Shale. " United Nations Conference on New Sources of Energy, E35-S83, Rome, 1961. New York: United Nations, 1964.

2048 Tamiya, H. "Growing Chlorella For Food and Feed." Proceedings World Symposium on Applied Solar Energy, Phoenix, AZ: 1955. pp. 231-42. Menlo Park, CA: Stanford Research Institute, 1956.

2049 Tamplin, Arthur R. Our Solar Energy Option--Physical and Biological. Livermore, CA: Lawrence Livermore Lab (University of California, Berkeley), 1973. 23pp. NTIS #UCRL-51315.

2050 _____ . "Solar Energy. " Environment 15(5): 16ff (8pp.), June 1973. Drawings, photos. , refs.

2051 Tan, H. M. , and Charters, W. W. S. "Experimental Investigation of Forced-Convection Heat Transfer. " Solar Energy 13(1):121-125, April 1970.

2052 Tanishita, I. "Present Situation of Commercial Solar Water Heaters in Japan. " Paper presented at Melbourne International Solar Energy Society Conference. 1970.

2053 _____ . "Recent Development of Solar Water Heaters in Japan. " United Nations Conference on New Sources of Energy, E35-S68, Rome, 1961. New York: United Nations, 1964.

2054 Tannenhaus, A. M. Inflated Rigidized Paraboloidal Solar Energy Concentrator Evaluation. Ger. 10295, Akron, OH: Goodyear Aircraft Corporation, July 1961.

2055 _____, et al. Solar Concentrator Systems for Space
Power Generator, Ger. 9711. Akron, OH: Goodyear Aircraft Corp., Feb. 1960.

2056 Tanner, C. B., Businger, J. A., and Kuhn, P. M.
"The Economical Net Radiometer." Journal of Geophysical Research 65: 3657, 1960.

2057 Tarnizhevskir, B. V., Savchenko, I. G., and Rodichev,
B. Y. "Results of an Investigation of a Solar Bat-
tery Power Plant." Geliotekhnika 2(2): 25-30, 1966.
(Russian)

2058 Teagan, W. P. "A Solar Powered Heating/Cooling
System with the Air Conditioning Unit Driven by an
Organic Rankine Cycle Engine." Proceedings of
the NSF/RANN Heating and Cooling for Buildings
Workshop, NSF/RANN-73-004, pp. 107-11, July
1973.

2059 "Tecnologia/Ingegneria Dell'Ambiente: Acqua Calda Dal
Sole." Casabella (text in English) No. 379: 54-57,
1973.

2060 Telkes, Maria. "Energy Storage Media." Proceedings
of the NSF/RANN Solar Heating and Cooling Buildings Workshop, NSF/RANN-73-004, pp. 57-59,
July 1973.

2061 _____. "Flat Tilted Solar Stills." Conference on
Solar and Aeolian Energy, Sounion, Greece, 1961.
New York: Plenum Press, 1964. pp. 14-18.

2062 _____. "Fresh Water from Sea Water by Solar
Distillation." Industrial Engineering Chemistry 45:
1108-14, 1953.

2063 _____. "Improved Solar Stills." Transactions of
the Conference on the Use of Solar Energy: The
Scientific Basis 3(2): 145-53. Tucson: University
of Arizona Press, 1958.

2064 _____. New and Improved Methods for Lower-Cost
Solar Distillation, PB 161, 402. Washington: U.S.
Department of Commerce Office of Technical Serv-
ices, 1959.

2065 _____. 'Nucleation of Supersaturated Salt Solutions. " Industrial & Engineering Chemistry 44: 1308+, 1952.

2066 _____. "A Review of Solar House Heating. " Heating and Ventilation pp. 68-74. Sept. 1949.

2067 _____. 'Review of Solar House Heating. " Heating and Ventilation 47: 72, Aug. 1950.

2068 _____. "Solar Cooking Ovens. " Solar Energy 3(1): 1-11, Jan. 1959.

2069 _____. "Solar Heat Storage. " Solar Energy Research (Daniels, F. and Duffie, J. A., eds.). Madison: University of Wisconsin Press, 1955. pp. 57-62.

2070 _____. "Solar Heat Storage. " ASME 64-WA/SOL-0, 1964.

2071 _____. "Solar House Heating, A Problem of Heat Storage. " Heating and Ventilating 44: 68, 1947.

2072 _____. "Solar Stills. " Proceedings, World Symposium on Applied Solar Energy, Phoenix, Arizona, 1955. Menlo Park, CA: Stanford Research Institute, 1956. pp. 73-78.

2073 _____. "Solar Stoves. " Transactions of the Conference on the Use of Solar Energy: The Scientific Basis 3: 87, 1958.

2074 _____. "Solar Thermoelectric Generators. " Solar Energy Research (Daniels, F. and Duffie, J. A., eds.) Madison: University of Wisconsin Press, 1955. pp. 137-42.

2075 _____. "Solar Thermoelectric Generators. " Transactions of Conference on the Use of Solar Energy: The Scientific Basis 5:1-7, Tucson: University of Arizona Press, 1958.

2076 _____, and Andrassy, S. "Practical Solar Cooking Ovens. " United Nations Conference on New Sources of Energy, E35-S101, Rome, 1961. New York: United Nations, 1964.

2077 _____, and Raymond, E. "Storing Solar Heat in Chemicals--A Report on the Dover House." Heating and Ventilating 80, 1949.

2078 Teplyakov, D. I. "Analytic Determination of the Optical Characteristics of Paraboloidal Solar Energy Concentrators." Geliotekhnika 7(5): 21-33, 1971 (in Russian).

2079 Thekaekara, M. P. "Data on Incident Solar Radiation." Supplement to Proceedings of the 20th Annual Meeting of the Institute for Environmental Science 21, 1974.

2080 _____. "The Solar Constant and Spectral Distribution of Solar Radiant Flux." Solar Energy 9(1):7-20, Jan. 1975.

2080a "Thermal Energy Creates Rotational Motion." Machine Design 46(23): 40, Sept. 19, 1974. Diagr.

2081 Thimann, K. V. "Solar Energy Utilization by Higher Plants." Proceedings, World Symposium on Applied Solar Energy, Phoenix, Arizona, 1955. Menlo Park, CA: Stanford Research Institute, 1956, pp. 255-59.

2082 "Thin Film in the Sun: Symposium on Solar Energy, Toulouse, France." Electronics 43: Sup 5E, July 20, 1970.

2083 Thirring, Hans. Energy for Man. New York: Harper & Row, 1958.

2083a "This Atrium Traps the Sun and It Cut a Six Week Heating Bill to $2.90." House and Home 46: 90-92, Aug. 1974.

2084 "This Building Saves Energy." Business Week (2305): 205-06, Nov. 10, 1973. Drawing.

2085 Thomason, Harry E. "Solar-Heated House Uses 3/4 HP for Air Conditioning." ASHRAE Journal (Journal of the American Society of Heating, Refrigerating, and Air Conditioning Engineers) Nov. 1962, pp. 58-62.

2086 _____ . Solar House Models. Barrington, NJ: Edmund Scientific, 1965. 35pp. Illus., plans, portraits.

2087 _____ . Solar Houses and Solar House Models. Available from Edmund Scientific Co., 150 Edscorp Building, Barrington, NJ: 08007. $1.00. 1972.

2088 _____ . "Solar Space Heating and Air Conditioning in the Thomason Home." Solar Energy Journal (4): 11-19, Oct. 1960.

2089 _____ . "Solar Space Heating, Water Heating, Cooling in the Thomason Home." United Nations Conference on New Sources of Energy, E35-S3, Rome, 1961. Also in Solar Energy 4: 11-19, 1960.

2090 _____ . "Three Solar Houses." Solar Energy 4(4): 11-19, Oct. 1960.

2091 _____ , and Thomason, Harry Jack Lee, Jr. Solar House Heating and Air-Conditioning Systems: Comparisons and Limitations. Barrington, NJ: Edmund Scientific Co., 1974. 61pp. Illus., refs.

2092 _____ , and _____ . "Solar House Plans." Avail. from Edmund Scientific, Barrington, NJ.

2093 _____ , and _____ . "Solar House/Heating and Cooling Progress Report." Solar Energy 15(1): 27-39, May 1973.

2094 Thomsen, D. E. "Farming the Sun's Energy." Science News 101(15): 237-38, April 8, 1972. Photos, diagrs.

2095 Threlkeld, J. L. Solar Energy as a Potential Heat Source for the Heat Pump. University of Minnesota #5566

2096 _____ , and Jordan, R. C. "Solar Collector Studies at the University of Minnesota." Transactions of the Conference on Use of Solar Energy: The Scientific Basis 2: 105-14.

2097 Thring, M. W. "Society for Social Responsibility in Science." Combustion 44(11): 19-21, May 1973.

2098 Tinker, Frank A. "Will Solar Farming Solve Our Power Crisis?" Popular Mechanics 138(1): 90, July 1972. Drawing, photo., diagr.

2099 Tonne, F. "Optico-Graphic Computation of Insolation-Duration and Insolation-Energy." Transactions of the Conference on the Use of Solar Energy: The Scientific Basis 1: 104-12, 1958.

2100 Touloukian, Y. S., et al. "Thermal Radiative Properties-Coatings." Thermophysical Properties of Matter. Plenum Data Corporation, Vol. 9, 1972.

2101 _____. "Thermal Radiative Properties-Metallic Elements and Alloys." Thermophysical Properties of Matter, Plenum Data Corporation, Vol. 7, 1970.

2102 _____. "Thermal Radiative Properties-Nonmetallic Solids." Thermophysical Properties of Matter. Plenum Data Corporation, Vol. 8, 1972.

2103 Towne School of Civil & Mechanical Engineering. Conservation & Better Utilization of Electric Power by Means of Thermal Energy Storage and Solar Heating. Oct. 1971. Distributed by NTIS (National Technical Information Service, U.S. Department of Commerce, 5285 Port Royal Rd., Springfield, VA 22151.

2104 Transactions of the Conference on the Use of Solar Energy: The Scientific Basis. Tucson: University of Arizona Press, 1958.

2105 Transactions of the NSF/NOAA Solar Energy Data Workshop. Washington, DC. Nov. 29-30, 1973.

2106 Trayser, D. A. and Eibling, D. A. "A 50-Watt Portable Generator Employing a Solar Powered Stirling Engine." Solar Energy 11(3-4):153-59, July 1967.

2107 Treble, F. C. "Space Power and the Solar Cell." Aeronautical Journal 74: 995-1000. Discussion 1000-1001, Dec. 1970.

2108 _____. Status Report on RAE (Royal Aircraft Establishment) Advanced Solar Array Development. Paper presented at the IEEE Photovoltaic Specialists' Conference, Silver Springs, MD, pp. 2-4, May 1972. NTIS #AD 755 743.

2109 Trinklein, F. E. "Solar Energy." The American Way 12-16, Feb. 1974.

2110 Trivich, D. "Photovoltaic Cells and Their Possible Use as Power Converters for Solar Energy." Ohio Journal of Science 53: 310-14, 1953.

2111 _____, Flinn, P. A., and Bowlden, H. J. "Photovoltaic Cells in Solar Energy Conversion." Solar Energy Research (Daniels, F. and Duffie, J. A., eds.). Madison: University of Wisconsin Press, 1955. pp. 148-54.

2112 Trombe, F. "Development of Large-Scale Solar Furnaces." Solar Energy Research (Daniels, F. and Duffie, J. A., eds.), p. 169-71. Madison: University of Wisconsin Press, 1955.

2113 _____. "Solar Furnaces and Their Applications." Solar Energy 1(2): 9-15, April 1957.

2114 _____. "Use of Solar Energy for High Temperature Processing (Solar Furnaces.), General Report." United Nations Conference on New Sources of Energy, E35-S Gr-S20, Rome, 1961. New York: United Nations, 1964.

2115 _____, and Foex, M. "Les Applications Practiques Actualles des Fours Solaires et Leurs Possibilitiés Economiques de Development." United Nations Conference on New Sources of Energy, E35-S, Rome, 1961. New York: United Nations, 1964.

2116 _____, and _____. "Economic Balance Sheet of Ice Manufacture with an Absorption Machine Utilizing the Sun as the Source of Heat." United Nations Conference on New Sources of Energy, E35-S, Rome, 1961. New York: United Nations, 1964.

2117 _____, and _____. "Production de Glace à l'Aide de l'Energie Solaire." Applications Thermiques de l'Energie Solaire dans le Domain de la Recherche et de l'Industrie, Symposium at Mont-Louis, France, 1958, Paris: Centre National de la

Recherche Scientifique, 1961, pp. 469-81.

2118 _____, and _____. "Production of Cold by
Means of Solar Radiation." Solar Energy 1(1): 51,
Jan. 1957.

2119 _____, and _____. "Traitements à Haute Temperature au Moyen de Fours Centrifugues Chauffes
par l'Energie Solaire." Transactions of the Conference on the Use of Solar Energy: The Scientific
Basis 2(1B): 146-86. Tucson: University of Arizona Press, 1958.

2120 _____, _____, and LaBlanchetais, C. H. "Conditions de Traitement et Mesures Physique dans
les Fours Solaires." United Nations Conference
on New Sources of Energy, E35-S, Rome, 1961.
New York: United Nations, 1964.

2121 _____, _____, and LePhat Vinh, M. "Research
on Selective Surfaces for Air Conditioning Dwellings." Proceedings of the U.N. Conference on
New Sources of Energy 4: 625, 638, 1964.

2122 _____, and LaBlanchetais, C. H. "Principles of
Air Conditioning in Countries with a Clear Sky."
United Nations Conference on New Sources of En-
ergy, E35-E111. New York: United Nations, 1964.

2123 Truhlar, E. J., et al. "Solar Radiation Measure-
ments in Canada. United Nations Conference on
New Sources of Energy, E35-S18, Rome, 1961.
New York: United Nations, 1964.

2124 "Turning on the Sunpower." Newsweek 82(3): 78, July
16, 1973. photo.

2125 Tybout, R. A., and Löf, G. O. G. "Solar House Heating." Natural Resources Journal 10(2): 283-326,
April 1970.

2126 TYCO Laboratories. Final Report #AFCRL-66-134,
1965.

2127 Tynes, A. R. "Integrating Cube Scattering Detector."
Applied Optics 9: 2706-10, Dec. 1970. Diagrs.

2128 United Nations. New Sources of Energy and Energy
      Development. Report on the United Nations Con-
      ference on New Sources of Energy: Solar Energy--
      Wind Power--Geothermal Energy, Rome, 21-31 Aug.
      1961. New York: United Nations, 1964. Vol. IV.
      Solar Energy (Electricity), 665pp. Illus. $7.50;
      Vol. V. Solar Energy II (Heating), 423pp. Illus.,
      $4.30; Vol. VI. Solar Energy III (Cooling, Distil-
      lation, Furnaces).

2129 _____. New Sources of Energy and Economic De-
      velopment: Solar Energy, Wind Energy, Tidal En-
      ergy, Geothermal Energy, and Thermal Energy of
      the Seas. New York: United Nations, 1957. Refs.

2130 _____. Solar Distillation as a Means of Meeting
      Small-Scale Water Demands. New York: United
      Nations. 1970. 86pp. Illus., plans (U.N. Doc.
      ST/ECA/121).

2131 _____. Department of Economic and Social Affairs.
      "Availability and Instruments for Measurements:
      Radiation Data, Networks, and Instrumentation."
      United Nations Conference on New Sources of En-
      ergy, Rome, 1961. New York: United Nations,
      1964. E/3577, rev. 1, Doc. ST/ECA72.

2132 _____. _____. Solar Distillation. Publication
      Sales No. E. 70. II. B. 1, 1970.

2133 _____. _____. Sources of Energy and Energy
      Development." United Nations Conference on New
      Sources of Energy, Rome, 1961. New York: United
      Nations, 1964. E/3577/rev. 1, Doc. ST/ECA/72.

2134 _____. Educational, Scientific, and Cultural Orga-
      nization (UNESCO). Wind and Solar Energy: Pro-
      ceedings of the New Delhi Symposium, Paris, 1956.
      128pp. Illus., maps, diagrs., tables. 57718.

2135 United States. Congress. House of Representatives.
      92nd Congress, Second Session. Solar Energy Re-
      search. Washington, DC: U.S. Government Print-
      ing Office, 1973. 119pp.

2136 _____. _____. _____. Committee on Sci-
      ence and Astronautics, 93rd Congress, First Ses-

sion. Solar Energy for Heating and Cooling. A
report on Hearings of June 7 and 12, 1973. 300pp.

2137 _____ . _____ . _____ . _____ . Solar Energy
for the Terrestrial Generation of Electricity. A
report on Hearings of June 5, 1973. 48pp.

2138 _____ . _____ . _____ . _____ . Solar Energy
Research, A Multi-disciplinary Approach. Staff Re-
port. Washington: U. S. Government Printing
Office, 1973. 119pp., Illus., refs. 73-601133.

2139 _____ . _____ . _____ . _____ . Solar Heating
and Cooling Demonstration Act. A report on Hearings of Nov. 13-15. 1973.

2140 _____ . _____ . _____ , 93rd Congress, Second Session. H. R. 11864. Solar Heating and Cooling Demonstration Act of 1974. Background and Legislative History, Feb. 1974. 306pp.

2141 _____ . _____ . Senate. Committee on Aeronautical and Space Sciences, 93rd Congress. Solar
Heating and Cooling, Hearings on S2658 and H. R.
-11864. Washington: U. S. Government Printing
Office, 1974. 274pp. Photos., charts, diagrs.

2142 _____ . _____ . _____ . Committee on Labor
and Public Welfare. Special Subcommittee on the
National Science Foundation. Solar Home Heating
and Cooling Demonstration Act, 1974 (Hearings).
Washington: Government Printing Office, 1974.
343pp. Photos., drawings, diagrs., charts, refs.

2143 _____ . Defense Documentation Center. Power
Supplies: An ASITA Report. Arlington, VA, 1961.
310pp. 61-64760.

2144 _____ . Library of Congress. Aerospace Technology Division. Direct Energy Conversion in the
U. S. S. R. Solar Cell Research; Comprehensive Report. Washington, 1963. 63-61140. Refs., illus.

2145 _____ . National Science Foundation. An Assessment of Solar Energy as a National Energy Resource. By NSF/NASA Solar Energy Panel.
Md: University of Maryland, Dec. 1972.

2146 University of Minnesota-Honeywell Report to NSF, Report NSF/RANN/SE/GI-34871/PR/73/2. "Research Applied to Solar-Thermal Power Systems."

2147 Usami, A. "Copper-doped Radiation-resistance n/p-type Silicon Solar Cells." Solid State Electronics 13: 1202-04, Aug. 1970. Refs.

2148 "Utilization de l'Energie Solaire." L'Architecture d'Aujourd'hui (167): 88-96, May 1973. Illus., plans, diagrs.

2149 "Utility Designs Energy-Saving Home." Electrical World 180: 84-85, Sept. 1, 1973. Diagrs.

2150 "Utility to Build Energy-conserving House." Engineering News-Record 190: 17, June 14, 1973.

2151 "Utility's Experimental Home to Use Solar-Heat Collector." Air Conditioning Heating and Refrigeration News 129: 1+, Aug. 13, 1973.

2152 Utz, J. A., and Braun, R. A. "Design and Initial Tests of a Stirling Engine for Solar Energy Applications." M. S. Thesis, Mechanical Engineering Department, University of Wisconsin, 1960.

2153 Van Cleave, D., and De Sanders, N. "Satellites to Relay Solar Energy for Earth Needs." Product Engineering 40(34): 13-14, Aug. 25, 1969. Illus.

2153a Van der Ziel, A. "Solar Power Generator with the Pyroelectric Effect." Journal of Applied Physics 45(9): 4128, Sept. 1974.

2154 Vasilev, A. M., and Landsman, A. P. Semiconductor Photoconverters. (Edited transcript of monography Poluprovodnikovye Fotopreobrazovateli, Moscow, 1971.) Trans. by Dean F. W. Koolbeck. 299pp. Reprint #FTD-HT-23-317-72. NTIS #AD-753 063.

2155 Versagi, F. J. "How Much of an Answer is Solar Energy?" Air Conditioning Heating and Refrigeration News 128(11): 25-26, March 12, 1973; 15-16, March 19, 1973.

2156 Vigren, D. D. "Solar Cooker is an Ecologist's De-

light. " <u>Popular Science</u> 199(2): 55, Aug. 1971.
Illus.

2157 Villecco, Marguerite. "Sunpower. " <u>Architecture Plus</u>
2(5): 84-99, Oct. 1974. Photos., diagrs., plans.

2158 Vincze, Stephen A. "A High-Speed Cylindrical Solar
Water Heater. " <u>Solar Energy</u> 13(3): 339-44, Nov.
1971.

2159 "Violet Cell Raises Solar Power Specs. " <u>Electronics</u>
45: 30-31, May 22, 1972.

2160 Vita Report 10. <u>Evaluation of Solar Cookers</u>, Publication and Technical Services Branch, U. S. Department of State 0-17, Washington, DC and Volunteers
for International Technical Assistance, Inc., Schenectady, NY.

2161 Walters, S. "Power in the Year 2001. " <u>Mechanical
Engineering</u> 93: 24-26, Sept. 1971. Refs., diagrs.
33-6, Nov. 1971.

2162 Walton, S. "World's First Solar-Powered Watch. "
<u>Popular Mechanics</u> 140: 172, Dec. 1973. Illus.

2163 "Wanted: A Coordinated, Coherent National Energy
Policy Geared to the Public Interest. " <u>Conserva-
tion Foundation Letter</u>, June 1972. 12pp.

2164 Ward, G. T. "Possibilities for the Utilization of
Solar Energy in Underdeveloped Rural Areas. "
Bulletin 16, Rome, Italy, Agricultural Engineering
Branch, World Food and Agricultural Organization
of the United Nations, 1961. pp. 5-6, and 19-21.

2165 Watson-Monro, C. N. "A New Look at Solar Radiation
as an Energy Source. " <u>Search</u> 4(4): 100-104, April
1973. Graphs, refs., table.

2166 Weaver, Kenneth F. "The Search for Tomorrow's
Power. " <u>National Geographic</u> 142(5): 650ff (32pp.),
Nov. 1972. Photos., diagrs.

2167 Weihe, Henrik. "Fresh Water from Sea Water: Distillation by Solar Energy. " <u>Solar Energy</u> 13(4):
439ff (5-1/2pp.), July 1972. Charts, diagrs.

2168 Weinberg, Alvin M. "Long-Range Approaches for Re-
solving the Energy Crisis." Mechanical Engineering
95(6): 14-19, June 1973. Diagrs., photo.

2169 Weinberger, H. "The Physics of the Solar Pond."
Solar Energy 8(2):45-46, April 1964.

2170 Weingart, Jerome. "Everything You've Always Wanted
to Know About Solar Energy, But Were Never
Charged Up Enough To Ask." Environmental Quality
3(9): 39ff 95pp., Dec. 1972. Chart, photos., diagr.

2171 _____. "Surviving the Energy Crunch." Environmental Quality 4(1): 28ff (7pp.), Jan. 1973. Photos.,
table.

2172 _____, and Schoen, Richard. "Project SAGE--An
Attempt to Catalyze Commercialization of Gas Sup-
plemented Solar Water Heating Systems for New
Apartments in Southern California." Proceedings
of the NSF/RANN Solar Heating and Cooling for
Buildings Workshop, NSF/RANN-73-004, pp. 75-91,
July 1973.

2173 _____, and Yasui, R. K. Stacked Solar Cell
Arrays (Patent). Commissioner of Patents,
Washington, DC 20231. Patent-3715600.

2174 Weinstein, A., and Chen, C. S. "Feasibility of Solar
Heating and Cooling of Buildings." Professional
Engineer 44: 28, 1974.

2175 "Welded Connections Brace Solar Cells for Longer Life;
Telsum Cell." Electronics 42: 183-84, Nov. 24,
1969. Illus.

2176 Wells, Malcolm B. "Confessions of a Gentle Archi-
tect." Environmental Quality 4(7): 51-58, July
1973. Diagrs., illus., table.

2177 Werner, R., and Ciarlariello, T. "Metal Hydride
Fuel Cells as Energy Storage Devices." United
Nations Conference on New Sources of Energy, E35-
Gen. 14, Rome, 1961. New York: United Nations,
1964.

2178 Wernick, S., and Pinner, R. Surface Treatment and

Finishing of Aluminum and Its Alloys. Teddington, England: Robert Draper, 1959.

2179 "What Future for Energy?" Nature 242(6): 357-59, April 6, 1973.

2180 Whillier, A. "Design Factors Influencing Solar Collectors." Low Temperature Engineering Applications of Solar Energy, New York, ASHRAE, 1967.

2181 _____. "Plastic Covers for Solar Collectors." Solar Energy 7(3):148-51, July 1963.

2182 _____. "Solar Radiation Graphs." Solar Energy 9(3):164-5, July 1965.

2183 _____. "Thermal Resistance of the Tube-Plate Bond in Solar Heat Collectors." Solar Energy 8(3): 95-98, July 1964.

2184 White, Gerald M. A Solar-Powered Air Conditioning System for Livestock Shelters, Purdue, #60-4223. 21/07/1882.

2185 "White House Science Report Stresses Coal, Solar Research Need." Coal Age 78: Jan. 1973.

2186 Whitlow, E. P., and Swearingen, J. S. "An Improved Absorption-Refrigeration Cycle." Paper presented at Southern Texas AIChE Meeting, 1959.

2187 Williams, D. A. "Intermittent Absorption Cooling Systems with Solar Regeneration." Refrigeration Engineering (66): 33, Nov. 1958.

2188 _____. "Selective Radiation Properties of Particulate Semiconductor Coatings on Metal Substrates." Ph.D. Thesis. Madison: University of Wisconsin, 1961.

2189 _____, Chung, R., Löf, G. O. G., Fester, D. A. and Duffie, J. A. "Cooling Systems Based on Solar Regeneration." Refrigeration Engineering 66: 33, Nov. 1958.

2190 _____, Lappin, T. A., and Duffie, J. A. "Selective Radiation Properties of Particulate Coatings."

American Society of Mechanical Engineers, Symposium, New York, 1962. Paper 62-WA-182.

2191 _____, Löf, G. O. G., Fester, D. A., and Duffie, J. A. "Intermittent Absorption Cooling Systems with Solar Regeneration." Refrigeration Engineering 66: 33, Nov. 1958.

2192 Williams, J. Richard. Solar Energy: Technology and Applications. Ann Arbor, MI: Ann Arbor Science Publishers, Inc., 1974. 119pp. Index, photos., graphs, refs. $9.95.

2193 Wilson, B. W. "The Role of Solar Energy in the Drying of Vine Fruits." United Nations Conference on New Sources of Energy, E35-S4, Rome, 1961. New York: United Nations, 1964.

2194 _____. "Solar Distillation in Australia." Transactions of the Conference on the Use of Solar Energy: The Scientific Basis 3: 154-58. Tucson: University of Arizona Press, 1958.

2195 Wilson, V. C. "The Conversion of Solar Energy to Electricity by Means of the Thermoionic Converter." United Nations Conference on New Sources of Energy, E35-S90, Rome, 1961. New York: United Nations, 1964.

2196 "Wind Generators Use Sun's Energy More Effectively Than Solar Cells." Electrical Review 192(22): 782-84, June 1, 1973.

2197 Witherall, P. G., and Faulhaber, M. E. "Silicon Solar Cell as a Photometric Detector." Applied Optics 9: 73-78, Jan. 1970. Diagrs., refs.

2198 Wohlrabe, Raymond. Exploring Solar Energy. New York: World Publishing Co., 1966. PLB $3.91. Illus. (Gr. 4-9). 93pp.

2199 Wolf, Martin. "Developments in Photovoltaic Solar Energy Conversion for Earth Surface Applications." United Nations Conference on New Sources of Energy, E35-S44, Rome, 1961. New York: United Nations, 1964.

2200 _____. Proceedings of the 9th IEEE Photovoltaic
Specialists Conference, Silver Spring, MD, May
1972.

2201 _____. "Solar Energy: An Endless Supply." Consulting Engineer 40(3): 171ff (9pp.), March 1973.
Diagr., graphs, table.

2202 _____. "Solar Energy Utilization by Physical
Methods." Science 184 (4134): 382-86, April 19,
1974. Photos., chart, drawing, graph.

2203 Woodmill, G. W. "The Energy Cycle of the Biosphere." Scientific American 233(3): 70+, Sept.
1970.

2204 World Symposium on Applied Solar Energy, Menlo
Park, CA, Association for Applied Energy Pro-
ceedings. New York: Johnson Reprint Corporation
Subs. of Academic Press, 1955. $15.00. Illus.

2205 Worstell, Charles C. A Solar Collector for Intermed-
iate Temperatures, University of Missouri. #66-
139. 26/09/5318.

2206 Wujek, J. H. "Silicon Solar Cells." Popular Electronics Including Electronics World 4: 98-102, Nov.
1973. Illus.

2207 "Yamashita's Mirrors on a Hot Tokyo Roof." UNESCO
Courier 27(1): 22-23, Jan. 1974. Photos.

2208 Yanagimachi, M. "How to Combine Solar Energy,
Nocturnal Radiation Cooling, Radiant Panel System
of Heating and Cooling, and Heat Pump to Make a
Complete Year Round Air-Conditioning System."
Transactions of the Conference on the Use of Solar
Energy: The Scientific Basis 3(2): 21-31. Tucson:
University of Arizona Press, 1958.

2209 _____. "Report on Two and One-Half Years Exper-
imental Living in Yanagimachi Solar House II.
"United Nations Conference on New Sources of
Energy, E35-S94, Rome, 1961. New York: United
Nations, 1964.

2210 Yellot, John I. "Calculation of Solar Heat Gain through

Single Glass. " <u>Solar Energy</u> 7(4): 167-75, 1963.

2211 _____. "Japan Applies Solar Energy. " <u>Mechanical Engineering</u> pp. 46-51, March 1969.

2212 _____. "The Measurement of Solar Radiation. " <u>Low Temperature Application of Solar Energy</u>, New York, American Society of Heat, Refrigeration, and Air Conditioning, 1967.

2213 _____. "Solar Energy Progress. " <u>Mechanical Engineering</u> 92: 28-34, July 1970. Refs., illus., diagrs.

2214 _____. "Solar Energy: Where It Will Shine in the Seventies. " <u>Chemical Engineering</u> 77: 85-89, June 29, 1970. Refs., illus., diagrs.

2215 _____. "Thin Film Water Heater. " <u>Solar and Aeolian Energy, Sounion, Greece,</u> 1961. New York: Plenum Press, 1964. pp. 112-23.

2216 _____. "Utilization of Sun and Sky Radiation for Heating and Cooling of Buildings. " <u>ASHRAE Journal</u> 15(12): 31-42, Dec. 1973. Refs., diagrs., graphs.

2217 _____, and Sobotka, R. "An Investigation of Solar Water Heater Performance. " <u>Trans ASHRAE</u> 70: 425-433, 1964.

2218 "Yocum Lodge: Warm-up. " <u>Progressive Architecture</u> 53(5): 84-87, May 1972. Illus., plans, diagrs.

2219 Young, G. J. <u>Fuel Cells.</u> New York: Reinhold, 1960, Vol. 2, 1963.

2220 Young, Patrick. "Seize That Sunbeam. " <u>National Observer</u> 1, July 8, 1972. Drawings, photo.

2221 Youngs, R. L. "Recommendation on Wood Drying. " <u>Forest Products Journal</u> 9(3): 121-24, 1959.

2222 Yuan, E. L., Evans, R. W., and Daniels, F. "Energy Efficiency of Photosynthesis by Chlorella. " <u>Biochemica Biophysica Acta</u> 17: 187-93, 1955.

2223 Zarem, A. M. and Erway, D. D., eds. Introduction
      to the Utilization of Solar Energy. New York:
      McGraw-Hill, 1963. 398pp. Illus. $13.50.

2224 _____. "Solar Sea Power." Physics Today 26:
      48-53, Jan. 1973 and reply by Marsh, H. S. 26:
      11+, June 1973.

2225 Zmuda, J. P. "Engine that Runs on Sunshine." Pop-
      ular Science 204(4): 87+, April 1974. Illus.

2226 Zwick, E. B. "Secondary Power." Space/Aero-
      nautics 52: 143-48ff., July 1969. Refs., illus.,
      diagrs.

3

GEOTHERMAL ENERGY

3000 Allen, Donard R., et al. "Earth's Heat Tapped for Geothermal Power Development." Public Power 29(6): 35-39, Nov.-Dec. 1971.

3001 _____. "Legal and Policy Aspects of Geothermal Resources Development." Water Resources Bulletin 8(2): 250ff (6-1/4pp.), April 1972.

3002 "The Alternatives: R&D for New Energy Sources." Economic Priorities Report 3(2): 23-26, May-June 1972. Graph.

3003 Altman, Manfred. "Conservation and Better Utilization of Electric Power by Means of Thermal Energy Storage and Solar Heating." NTIS Report PB-210 359, Oct. 1971. 263pp.

3004 "America's Energy Crisis." Newsweek 81(4): 52ff (5pp.), Jan. 22, 1973. Photos., tables.

3005 Anderson, Richard J. "The Promise of Unconventional Energy Sources." Battelle Research Outlook 4(1): 22ff (4pp.), 1972. Photos.

3006 "Arizona Geothermal Well Hits Steam." Oil & Gas Journal 71(16): 147, April 16, 1973.

3007 Armstead, H. C. H. Geothermal Energy: Review of Research and Development. Paris: UNESCO, 1973. 186pp. $14.00.

3008 Armstrong, Ellis L. "The Roar from an Emerging Source." Reclamation Era 57(3): 1ff (7pp.), Aug. 1971. Diagr., photos.

3009 "Auctioning the Rights to Geothermal Energy." Business Week (2315): 22, Jan. 26, 1974.

3010 Barnes, Joseph. "Geothermal Power." Scientific
      American 226(1): 70-77, Jan. 1972. Photos.,
      diagrs.

3011 _____. "Multipurpose Exploration and Development
      of Geothermal Resources." Natural Resources
      Forum 1(1): 55-58, 1971. New York: United Na-
      tions, $2.00.

3012 Barnes, Peter. "Escaping Steam." New Republic
      168(15): 9, April 14, 1973.

3013 Bienstock, D., et al. "Corrosion of Heat Exchange
      Tubes in a Simulated Coal-fired MHD System."
      Journal of Engineering for Power 93A(2): 249ff
      (8pp.), April 1971. (Pittsburgh Energy Research
      Center, research report.) Charts, diagrs., photos.

3014 "Big Plans for a Hot Resource." Chemical Week 108
      (20): 29, May 19, 1971. Illus.

3015 Bowen, Richard G. "Environmental Impact of Geo-
      thermal Development." Aware 32(7): May 1973.

3016 _____. "Lessons Learned in Steam Exploration."
      Oil & Gas Journal 69: 174-5, March 22, 1971. Map.

3017 _____, and Groh, Edward A. "Geothermal-Earth's
      Primordial Energy. Technology Review 74(1): 42ff
      (7pp.), Oct.-Nov. 1971. Charts, diagr.

3018 Bradbury, J. C. C. "The Economics of Geothermal
      Power." Natural Resources Forum 1(1): 46-53,
      1971. New York: United Nations, $2.00.

3019 "Briefings Before the Task Force on Energy--Vol. 3."
      Report, House Committee Science Astronautics 92
      Con II Serial U, May 1972. 204pp.

3020 Briggs, Peter. "From Dante's Inferno Power for
      Millions." International Wildlife 2(4): 14-16, July-
      Aug. 1973. Drawings.

3021 Buckman, Robert D. "The Major Energy Market."
      American Gas Association Monthly 54(9): 4ff (4pp.),
      Sept. 1972. Photos.

3022 Cady, C. V., et al. "Model Studies of Geothermal
     Steam Production." AlChE Water Symposium 69:
     439-45, 1972. Diagr.

3023 Carlsmith, R. S. Alternative Sources of Energy.
     Paper presented at the 26th Annual Conference of
     the Middle East Institute, Washington, DC, Sept.
     29, 1972. NTIS # Conf-720963-1.

3023a Carter, L. J. "Solar and Geothermal Energy: New
     Competition for the Atom." Science 186(4166):
     811-13, Nov. 29, 1974.

3024 Clark, Wilson, et al. "How to Kick the Fossil Fuel
     Habit." Environmental Action 5(18): 3-6, 12-15,
     Feb. 2, 1974. Drawing.

3025 "Combine Wants Geothermal 'In'." Electrical World
     176(12): 31, Oct. 1, 1971.

3026 "Court Okays More Steam." Chemical Week 110: 22,
     June 21, 1972.

3027 Cowan, Edward. "Geothermal Power Hunted in Mon-
     tana." New York Times, June 9, 1974.

3028 "Critics Hit Geothermal Guidelines." Electrical World
     176(7): 32, Dec. 15, 1971.

3029 Cromling, John. "Geothermal Drilling in California."
     Petroleum Technology 25: 103ff (6pp.), Sept. 1973.

3030 _____. "How Geothermal Wells are Drilled and
     Completed." World Oil 177(7): 42-48, Dec. 1973.
     Diagrs., photo.

3031 Denton, Jesse C. and Dunlop, Donald D. "Geothermal
     Resources Research." Aware (39): 13-15, Dec.
     1973.

3032 De Onis, Juan, et al. "Energy Crisis: Shortages and
     New Sources." New York Times p. 1, April 16,
     1973; p. 1, April 17, 1973; p. 1, April 18, 1973.
     Diagr., graph, map, photos.

3033 Dixon, Rod P. Recovery of Geothermal Energy by
     Means of Underground Nuclear Detonations. Wash-

ington: Commissioner of Patents, filed June 1970,
patented Feb. 1972. 3pp.

3034 "Drilling Probes Geothermal Sources." Oil & Gas
Journal 71(24): 100, June 11, 1973. Photo.

3035 Driscoll, James G. "A Few Victories in the War for
Clean Power." National Observer p. 4, Sept. 6,
1971. Illus.

3036 "Dry Geothermal Areas Have Huge Power Potential."
Engineering News-Record 188: 11, June 29, 1972.

3037 "Dry Geothermal Wells: Promising Experimental Results." Science 182(4107): 43-45, Oct. 5, 1973.
Diagr.

3038 Dvorov, I. M. "Deep Seated Heat from the Earth."
Glubiwnoe Teplo Zemli (Russian) p. 9-104, 141-
205, 1972.

3039 "Energy Crisis: Are We Running Out?" Time 99(24):
49-53, June 12, 1972.

3040 "Energy from Above and Below." Nature 242(5392):
6-8, March 2, 1973.

3041 "Energy in the Year 2000." Power from Oil p. 3-6,
Fall 1973. Photos.

3042 "Energy: More Geothermal Power." Chemical and
Engineering News 48: 12-13, Aug. 17, 1970. Map.

3043 "Energy Options for the United States." IEEE Spectrum 10(9): 63ff (6pp.), Sept. 1973. Graphs, table.

3044 Everitts, D. J. "The Role of State Government in the
Development of Geothermal Resources." AIChE
Water Symposium Series 59: 459-65, 1972. Photos.,
diagr.

3045 Fanin, Senator Paul. "Law Needed for Geothermal
Energy?" Public Utilities Fortnightly 93(3): 15-18,
Jan. 31, 1974.

3046 Fenner, David and Klarmann, Joseph. "Power from
the Earth." Environment 13(10): 19ff (12pp.), Dec.

1971. Photos., diagrs.

3047 Finney, John P. "Heat Transfer: The Geysers Geothermal Power Plant." Chemical Engineering Progress 68(7): 83ff (4pp.), July 1972. Chart, photos.

3048 _____. "Practical Application of Geothermal
Steam." Public Utilities Fortnightly 93(3): 18-21,
Jan. 31, 1974.

3049 "First Non-California Geothermal Well is Sought."
Oil & Gas Journal 71(9): 90, Feb. 26, 1973.

3050 "Fluid Power Sleuths Intensify Search for Ecological
Villains: Clean Power Has No Smoke." Product
Engineering 41: 76ff, Dec. 7, 1970.

3051 Friedlander, Gordon D. "Energy: Crisis and Challenge." IEEE Spectrum 10(5): 18ff (10pp.), May
1973. Diagrs., graphs, photo., table.

3052 Friz, Thomas O. "Geothermal As a Future Energy
Source--By Hydraulic Fracturing and Nuclear Ex-
plosive." Aware 37: 13-15, Oct. 1973. Table.

3053 _____. "Three Major Types of Geothermal Sys-
tems--Brine, Hot Dry Rock, 'Hot Spot.'" Aware
34(8): 8-11, July 1973. Table.

3054 Fuchs, Rober L., and Westphal, Warren H. "Energy
Shortage Stimulates Geothermal Exploration."
World Oil 177(7): 37-42, Dec. 1973. Photos.,
tables.

3055 "Furnace Below the Crust." New York Times Sec. 8,
p. 8, Nov. 5, 1972. Photo.

3056 Gannon, Robert. "Atomic Power, What Are Our Alternatives?" Science Digest 70(6): 18ff (5pp.), Dec.
1971. Photos.

3057 Garrison, L. E. "Geothermal Steam in the Geysers-
Clear Lake Region California." Bulletin of the
Geological Society of America 83: 1449-68, May
1972. Maps, refs.

3058 "Geothermal Activity Heating Up in West." Electrical

World 175: 25-26, Jan. 15, 1971.

3059 "Geothermal Energy Comes Clean. " Chemical Week
113(25): 37, Photo.

3060 "Geothermal Energy for Electric Power. " Atom (U. S.)
8(10): 10-14, Dec. 1971. Map, photo. , diagrs.

3061 "Geothermal Energy Starts to Steam. " Industrial Research 15: 24-25, March 1973.

3062 "Geothermal In Iceland. " Power Engineering 76: 38-
39, March 1972.

3063 "Geothermal Power. " Mechanical Engineering 94: 32-
33, March 1972.

3064 "Geothermal Power Is Aim of New Plowshare Project."
Industrial Research 12: 22, July 1970. Illus.

3065 "Geothermal Power: The Great Land Rush of '73. "
Forbes 111(2): 31, Jan. 15, 1973. Map.

3066 "Geothermal Project in Imperial Valley to Be Joint
Effort. " Oil & Gas Journal 70(47): 118, Nov. 13,
1972.

3067 "Geothermal Resources. " Mechanical Engineering 93:
51, May 1971.

3068 "Geothermal Resources Gather a Head of Steam. "
Engineering News-Record 186(18): 30ff (4pp.), May
6, 1971. Diagrs. , photos.

3069 "Geothermal Steam: Deepest Gates to Hell. " Chemical
and Engineering News 49: 9, Aug. 23, 1971.

3070 Gilmore, C. P. "Hot New Prospects for Power from
the Earth. " Popular Science 201(2): 56-59, Aug.
1972. Drawings, photos.

3071 Godwin, L. H. , et al. "Classification of Public Lands
Valuable for Geothermal Steam and Associated Geothermal Resources. " USGS Circular 647, USGPO,
1971. 18pp. Maps, charts. (Avail. in Micro-
fiche.)

3072 Goldsmith, M. "Geothermal Resources in California: Potentials and Problems." Avail. from Caltech Environmental Quality Laboratory, 1201 E. California Blvd., Pasadena, CA 91109.

3073 Gottschalk, Earl C., Jr. "Steam Below Ground Seen Giving Big Boost to U.S. Energy Supply." Wall Street Journal, p. 1, March 20, 1973.

3074 Gould, William R. "Energy Options for the Future." Speech presented at National Science Teachers Association Convention, Detroit, April 2, 1973. 13pp.

3075 "Great Land Rush of '73." Forbes 111(2): 31, Jan. 15, 1973. Map.

3076 Grey, Jerry. "Prospecting for Energy." Astronautics & Aeronautics 11(8): 24ff (5pp.), Aug. 1973. Drawings, refs.

3077 Guiles, R. A. "Geysers Erupt as Important Power Source." Iron Age 211(22): 32, May 31, 1973. Photo.

3078 Hamilton, Rep. Lee H. "The Energy Crisis: The Persian Gulf." Vital Speeches 39(8): 226ff (5pp.), Feb. 1, 1973.

3079 Hammond, Allen L. "Energy Options: Challenge for the Future." Science 177(4052): 875-77, 1972. Charts.

3080 _____. "Geothermal Energy: An Emerging Major Resource." Science 177(4053): 978-81, Sept. 15, 1972. Photo.

3081 Hauser, L. G. "Assessing Advanced Methods of Generation." Electrical Review 193(7): 219ff (4pp.), Aug. 17, 1973. Diagrs.

3082 Hayeland, James A. "Hellpower." Ecology Today 2(1): 36-40, March-April 1972. Photos.

3083 Hebb, David H. "Some Economic Factors of Geothermal Energy." Mines Magazine 62(7): 15ff (4pp.), July 1972. Charts, graph, diagrs., refs.

3084 Heldren, John. "Defusing Old Smokey by Plugging into
       Nature." Sierra Club Bulletin 56(8): 24-28, Sept.
       1971. Drawing.

3085 Henahan, John. "Full Steam Ahead for Geothermal
       Energy." New Scientist 57(827): 16-18, Jan. 4,
       1972. Map, photo.

3086 Heronemus, William E. "Alternatives to Nuclear En-
       ergy." Catalyst for Environmental Quality 2(3):
       21-25, 1972. Drawing, photo.

3087 "Hot Springs Fuel Iceland's Hopes." Business Week
       (2171): 124, April 10, 1971.

3088 "House, Senate Pass Geothermal Energy Bills." Con-
       gressional Quarterly 32(29): 1900, July 20, 1974.

3089 Hubbert, M. King. "Energy Resources: A Report to
       the Committee on Natural Resources of the National
       Academy of Sciences." National Research Council,
       Washington, DC: NAS--National Research Council,
       1962. 153pp.

3090 _____. "Energy Resources for Power Production."
       USGS, Paper presented at IAEA/USAEC Conference
       on Environmental Aspects of Nuclear Power Sta-
       tions, New York City, Aug. 10-14, 1970. 30-1/2pp.
       Charts, diagr. (Available from Environmental
       Access.)

3091 _____. "The Energy Resources of the Earth."
       Scientific American 224(3): 60ff (11pp.), Sept. 1971,
       Charts, diagrs., photo.

3092 "Improving Sophisticated Energy Sources May Answer
       Power Problems." Commerce Today 3(9): 4ff
       (5pp.), Feb. 5, 1973. Drawings, photos.

3093 "In Committee: AEC Authorization." (Operation Plow-
       share) Congressional Quarterly 32(15): 951-52,
       April 13, 1974.

3094 "In Committee: Geothermal Energy." Congressional
       Quarterly 32 (19): 1228, May 11, 1975.

3095 "In Committee: Geothermal Energy." Congressional

Quarterly 32(21): 1392, May 25, 1974.

3096 "In Committee: Geothermal Energy." Congressional
Quarterly 32(26): 1701, June 29, 1974.

3097 "In Committee: National Science Foundation." Congressional Quarterly 32(17): 1047-48, April 27, 1974.

3098 "In Committee: Related Hearings: Geothermal Energy."
Congressional Quarterly 32(6): 313, Feb. 9, 1974.

3099 "Industry Gets More Active in Geothermal Arena."
Oil & Gas Journal 72(17): 88, April 29, 1974.

3100 Ingraham, Norman P. "California Municipal Group
Studies Geothermal Power." Public Power 31(5):
28, Sept. -Oct. 1973.

3101 "Interior to Implement Geothermal Steam Act." Public
Utilities Fortnightly 87(3): 42, Feb. 4, 1971.

3102 "Interior's Proposed Geothermal Guidelines Please No-
body." Electrical World 177: 27, Feb. 15, 1972.

3103 International Society for Geothermal Engineering, Inc.
P. O. Drawer 4743, Whittier, CA: 90607.

3104 "Joint Venture to Tap the Earth's Heat." Business
Week 53, Nov. 11, 1972.

3105 Kaufman, Alvin. "An Economic Appraisal of Geothermal Energy." Public Utilities Fortnightly 88
(7): 19ff (5pp.), Sept. 30, 1971.

3106 Keene, Jenness and Arden, Tom. "Geothermal Stations
Have Pollution Problems, Too." Power 115(5): 96,
May 1971. Photo.

3107 Keller, G. V. and Romig, P. R. "Energy--A National Need." Mines Magazine 63(5): 14-17, May
1973.

3108 Kemesis, Paul. "NATO Advancing on Environment."
New York Times Oct. 28, 1973. Sec. 1, p. 8.

3109 Kent, M. F. "Power Generation Options for the
Eighties and Nineties." Public Utilities Fortnightly

90(4): 17ff (5-1/2pp.), Aug. 17, 1972. Drawings.

3110 Klaff, Jerome L. "Power Policies." Public Utilities Fortnightly 92(6): 31-34, Sept. 13, 1973.

3111 Kruger, Paul. "The Energy Outlook: Now to 1985." Aware (36): 3-8, Sept. 1973. Graph, tables.

3112 _____, and Otte, Carel, eds. Geothermal Energy: Resources, Production, Stimulation. Stanford, CA: Stanford University Press, 1973. 360pp. Diagrs., photos., graphs, index. $17.50.

3113 Kuwada, J. T. "Geothermal Power Plant Design." AlChE Water Symposium Series 69: 439-45, 1972. Diagr.

3114 _____, and Ramey, Henry Jr. "The Challenge of Geothermal Energy." Energy Environment Productivity Proceedings of the First Symposium on RANN (Research Applied to National Needs), Washington, DC, Nov. 18-20, 1973. Washington: National Science Foundation, pp. 33-38. Illus.

3115 Laird, Alan D. K. "The Emerging Geothermal Power Source." Technical Report presented at the American Geophysical Union Symposium, San Francisco, Dec. 6-9, 1971. 9pp. Diagr.

3116 _____. "The Emerging Geothermal Power Source." Paper presented at the 52nd Annual American Geophysical Union Meeting, Washington, DC, April 12-16, 1971. 9pp. Diagrs. (Avail. from Environment Information Access #72-08236.)

3117 "Let's Get Into Hot Water." Industrial Research 15: 22, Aug. 1973.

3118 "Lighting Some Candles." Architectural Forum 139(1): 89ff (10pp.), July-Aug. 1973. Drawing, photos.

3119 Loehwing, D. A. "Nature's Teakettle: Geothermal Power is Getting Up a Head of Steam." Barron's 53: 3+, Aug. 13, 1973. Illus.

3120 Lovins, Amory B. "The Case Against the Fast Breeder Reactor." Bulletin of Atomic Scientists

140 / Alternate Sources of Energy

29(3): 29ff (7pp.), March 1973. Refs.

3121 McCallum, John and Faust, Charles L. "New Frontiers in Energy Storage." Battelle Research Outlook 4(1): 26-30, 1972. Chart, photo.

3122 McCaslin, John C. "Geothermal Energy Has Vast
Global Potential." Oil & Gas Journal 71(12): 99,
March 19, 1973.

3123 McCormack, Rep. Mike. "Energy Crisis--An In-Depth
View." Chemical & Engineering News 51(14): 1-3,
April 2, 1973.

3124 _____. "We Must Explore New Methods of Energy
Conversion." Public Power 31(1): 20ff (4pp.),
Jan.-Feb. 1973. Diagr.

3125 MacDonald, Gordon A. Geological Prospects for Development of Geothermal Energy." Pacific Science
27(3): 209ff (11pp.) July 1973. Map, refs., tables.

3126 Mahon, William A. J., and Finlayson, James B. "The
Chemistry of the Broadlands Geothermal Area, New
Zealand." American Journal of Science 272(1): 46-
68, Jan. 1972. Charts, map, diagrs.

3127 Malin, H. Martin, Jr. "Geothermal Heats Up." Environmental Science and Technology 7(8): 680-82,
Aug. 1973. Map, photos.

3128 Manley, C. Conrad. "Mexico Uses Steam from the
Earth to Provide Lights for Its Cities." Christian
Science Monitor May 26, 1973. Map.

3129 Marinelli, Giorgio. "Deep Down Power." Development Forum 1(3): 5ff (5pp.), April 1973. Map.

3130 Matthew, Paul. "Geothermal Operating Experience at
Geysers Power Plant." Journal of Power Division
ASCE 99(2): 329ff (10pp.), Nov. 1973. Diagr.,
map, photo., tables.

3131 Meidav, T. "Geothermal Opportunities Bear Close
Look." Oil & Gas Journal 72(19): 102-06, May 13,
1974. Refs., map. graphs, chart.

3132 Miller, L. B. "Energy Outlook Grim Now, But May Ease by 1985." Industrial Development 142(2): 10-13, March-April 1973. Drawing.

3133 "Mobile Rig Drills Geothermal Well in Chile." Electrical World 174: 26, Sept. 1, 1970.

3134 Muffler, L. J. P., and White, D. E. "Geothermal Energy." Science Teacher 39(3): 40-44, March 1972. Map, chart, diagrs.

3135 [no entry]

3136 Nash, Barbara. "Energy from the Earth and Beyond." Chemistry 46(9): 6ff (4pp.), Oct. 1973. Diagrs., map, photo.

3137 "Natural Steam Can Produce Power Cleanly." Prevention 23(2): 116-17, Feb. 1971.

3138 "New Harness for Energy." Chemical Week 112(6): 45-47, Feb. 7, 1973. Photo.

3139 "New Waters to Rescue Southwest?" Christian Science Monitor April 8, 1971. p. 5.

3140 Nice, Joan. "Earth Energy." High Country News 6 (19): 1ff (3-1/2pp.), Sept. 27, 1974. Photos., map.

3141 Nixon, Richard M. "Message from the President of the U.S. Concerning Energy Resources." House Document 93-85, April 18, 1973. 17pp.

3142 "Ocean/Thermal: An Unmixed Blessing." Mosaic 5(2): 14+, Spring, 1974.

3143 O'Connor, J. J. "What Can We Expect from Geothermal?" Power 117(2): 9, Feb. 1973.

3144 O'Keefe, William. "Geothermal Power: Sleeping Giant Stirs, But Will Require Years to Waken Fully." Power 117(4): 32-35, April 1973. Diagrs.

3145 "On the Floor: Geothermal Energy." Congressional Quarterly 32(35): 2395-96, Aug. 31, 1974.

3146 "On the Floor: National Science Foundation." Congressional Quarterly 32(18): 1132-33, May 4, 1974.

3147 "On the Floor: Public Works--AEC Appropriations."
     (Plowshare) Congressional Quarterly 31(27): 1841,
     July 7, 1973.

3148 O'Neill, J. J., et al. The Development of a Resi-
     dential Heating and Cooling System Using NASA De-
     rived Technology. Huntsville, AL: Lockheed Re-
     search and Engineering Center, 1972. 99pp. NTIS
     # N73-17911.

3149 Papin, Warren J. "Steam Cleaned Water." Water
     Spectrum 3(4) 37-41, Winter 1971-72. Photos.

3150 Peet, Creighton. "Ultima Thule." American Forests
     77(6): 12-15, June 1971.

3151 Pickering, William H. "Important to Examine Possi-
     bilities of Unconventional Sources." Aware (31):
     5-7, April 1973.

3152 Plumlee, R. H. Perspectives in U.S. Energy Re-
     source Development. Albuquerque, NM: Sandia
     Labs, 1973. 113pp. NTIS # SLA 73-163/GA.

3153 Porter, Lloyd R. "Geothermal Resource Investiga-
     tions." Journal of Hydraulics Division ASCE
     (American Society of Civil Engineers) 99(11): 2097ff
     (15pp.), Nov. 1973. Diagrs., graphs, maps,
     photos., table.

3154 "Potential for Geothermal Power." Business Week
     74-75, March 17, 1973.

3155 "Power from a Hot Ocean." Mechanical Engineering
     92:53, Oct. 1970.

3156 "Power from the Earth." The Economist 238: 37,
     Jan. 23, 1971. Illus.

3157 "Pressure Builds for Geothermal Steam." Industry
     Week 178: 52, Aug. 27, 1973. Illus.

3158 "Priority Projects Show Increase in FY '72 Budget for
     Reclamation." Water Life 35(4): 1, April 1971.

3159 "Problems Beneath the Surface." Mosaic 5(2): 14+,
      Spring 1974.

3160 Quigg, P. W. "Alternative Energy Sources." Satur-
      day Review/World 1(12): 29-30, Feb. 23, 1974.

3161 Radin, Alex. "How We Can Conserve Energy."
      American Public Power Association 31(4): 6-9,
      July-Aug. 1973.

3162 "Report on a Plowshare Geothermal Power Plant."
      Nuclear News 14(7): 33, July 1971. Diagr.

3163 Rex, Robert W. "Geothermal Energy--the Neglected
      Energy Option." Atomic Scientists 27(8): 52ff
      (5pp.), Oct. 1971.

3164 Risser, Hubert E. "The U. S. Energy Dilemma: The
      Gap Between Today's Requirements and Tomor-
      row's." Illinois State Geological Survey Report 64.
      July 1973. 59pp.

3165 Ritter, William W. "Geothermal Energy: Prospects
      and Problems." Journal of Environmental Health
      35(5): 432ff (4pp.), March-April 1973. Photo.,
      tables.

3166 Robson, Geoffrey. "Geothermal Electricity Produc-
      tion." Science 184(4134): 371-75, April 19, 1974.
      Photos., drawing, refs., chart.

3167 Sandquist, Gary M., and Whan, Glenn A. "Environ-
      mental Aspects of Nuclear Geothermal Heat Stimu-
      lation." Aware 33(10): 10ff (5pp.), June 1973.

3168 Schactman, Tom. "Getting More Power to the People."
      Ecology Today 1(7): 46ff (3-1/2pp.), Sept. 1971.
      Photos.

3169 Schuster, Ray. "Turning Turbines with Geothermal
      Steam." Power Engineering 76(3): 36-41, March
      1972. Photos.

3170 Scott, Stanley, and Wood, Samuel E. "California's
      Bright Geothermal Future." Cry California 7(1):
      10ff (14pp.), Winter 1971-72. Maps, drawings,
      photos.

3171 Seaborg, Glen T., et al. "Nuclear and Other Energy."
NY Academy of Science Annals 216(79): 79ff (10pp.),
May 18, 1973.

3172 "Secretary Morton Approves Geothermal Development."
USDI News release. Dec. 18, 1973. 14pp. Map.

3173 "Shell High Bidder in Geothermal Sale." Oil & Gas
Journal 72(10): 42, March 11, 1974.

3174 Small, HuDee. Nature's Tea Kettle. Avail. from
Geothermal Information Services, 318 Cherrywood
St., West Covina, CA 91791. $9.95.

3175 Smil, Vaclav. "Energy and the Environment." Futurist 8(1): 4-13, Feb. 1974. Photos., diagrs.

3176 Smith, M. C. Geothermal Energy. Los Alamos, NM:
Los Alamos Scientific Lab, 1973. 31pp. NTIS
#LA-5289-MS/GA.

3177 Smith, William D. "The Geyser: An Unusual Source
of Energy." New York Times p. 2, Jan. 14, 1973.
Photo.

3178 Snyder, M. Jack, and Chilton, Cecil. "Planning for
Uncertainty: Energy in the Years 1975-2000."
Battelle Research Outlook 4(1): 2-6, 1972. Graphs.

3179 "Solar Geothermal Energy Agencies: Are They
Needed?" Congressional Quarterly 32(20): 1227-28,
May 11, 1974.

3180 Spurgeon, David. "Nuclear Power for the Third
World." New Scientist 60(875): 694-87, Dec. 6,
1973.

3181 Starr, Chauncey. "Research on Future Energy Systems." Professional Engineer 42(2): 20ff (4-
1/2pp.), Feb. 1972. Chart, drawings.

3182 "Starting to Tap the Hot Reservoirs." Mosaic 5(2):
24+, Spring 1974.

3183 "The Steam Inside." Newsweek 77(23): 95, June 7,
1971. Photo.

3184 Stevovich, V. A. Soviet Geothermal Electric Power
      Engineering Report 2. Rockville, MD: Informatics
      Inc., 1972. 84pp. NTIS #AD-754 947.

3185 "Subterranean Ocean Could Solve Power Problem."
      Engineering News-Record 185: 16, Aug. 13, 1970.

3186 Sullivan, Walter. "Using the Atom to Run a Steam
      Engine." New York Times, Aug. 25, 1971.

3187 "Taking Mother Earth's Temperature." Mosaic 5(2):
      24+, Spring 1974.

3188 Tazieff, Haroun. "Ethiopia's Geothermal Possibili-
      ties." Compressed Air Magazine 77(1): 14-17,
      Jan. 1972. Map, chart, photos.

3189 "Team To Dig Steam." Chemical Week 111: 24, Nov.
      8, 1972.

3190 "There's A Lot of Steam Under Them Thar Hills."
      Electrical World 174: 26, Sept. 1, 1970. Illus.,
      map.

3191 United Nations. New Sources of Energy and Economic
      Development: Solar Energy, Wind Energy, Tidal
      Energy, Geothermal Energy, and Thermal Energy
      of the Seas. New York: United Nations, 1957.
      Refs.

3192 _____. New Sources of Energy and Energy De-
      velopment. Report on the United Nations Confer-
      ence on New Sources of Energy: Solar Energy,
      Wind Power, Geothermal Energy. Rome, 21-31
      Aug. 1962. 65pp. New York: United Nations,
      1962. $0.75.

3193 _____. Proceedings of the United Nations Confer-
      ence on New Sources of Energy: Solar Energy,
      Wind Power, and Geothermal Energy, Rome 21-31
      Aug. 1961. New York: United Nations, 1964. Vol.
      II Geothermal Energy I, 420pp. Illus., $5.00; Vol.
      III Geothermal Energy II, 516pp. Illus.

3194 "United Nations Symposium on the Development and
      Utilization of Geothermal Resources, Pisa, Italy.
      Sept. 22-Oct. 1: Preview." Electrical World 173:

32-34, June 22, 1970. Illus. , diagr.

3195 United States. Congress. House of Representatives.
Committee on Science and Astronautics. Subcomit-
tee on Energy. Geothermal Energy--Resource and
Technology. Washington: U. S. Department of the
Interior, Sept. 13, 1973. 22pp.

3196 _____. Department of the Interior. "Revised Geothermal Leasing Proposal Announced. " News release, 1pp. , Nov. 29, 1972.

3197 "Utilities Put Faith in Old Faithful. " Business Week
(2144): 50-52, Oct. 3, 1970. Diagr.

3198 Weaver, Kenneth F. "The Search for Tomorrow's
Power. " National Geographic 142(5): 650ff (32pp.),

3199 Wehlage, Edward F. "Tests Run on New Prime Mover
for Geothermal Power Generation. " Consulting
Engineer 41(2): 128, Aug. 1973.

3200 Weinberg, Alvin M. "Long-Range Approaches for Re-
solving the Energy Crisis. " Mechanical Engineering 95(6): 14ff (5pp.) June 1973. Diagrs. , photo.

3201 Weismantel, G. E. "Geothermal Power Perks Up. "
Chemical Engineering 77: 24-25, Nov. 30, 1970.
Illus. , diagrs.

3202 _____. "Geothermal Power Still Iffy. " Chemical
Engineering 80(6): 40-42, March 5, 1973. Diagrs. ,
photo.

3203 Weissberg, B. G. , and Zobel, M. G. R. "Geothermal
Mercury Pollution in New Zealand. " B Env Contam
Tox 9(3): 148ff (8pp.), March 1973. Graph, map,
tables.

3204 Werth, Glen C. "Nuclear Explosives Applications. "
Nuclear News 14(10): 103-05, Oct. 15, 1971.
Chart, photos.

3205 "Western States Rich in Geothermal Potential. " Engineering and Mining Journal 172 32, Dec. 1971.

3206 "What Future for Energy?" Nature 242(6): 357-59,

April 6, 1973.

3207 White, D. E., et al. "Vapor-dominated Hydrothermal Systems Compared with Hot-water Systems." Economic Geology and the Bulletin of the Society of Economic Geologists 66: 75-97, Jan. 1971. Maps, diagrs., refs.

3208 Wilford, John Noble. "Geothermal Energy Held a Vast World Reservoir." New York Times p. 42, Jan. 11, 1973.

3209 _____. "New Plan Is Outlined for Tapping Geothermal Energy." New York Times p. 18, June 21, 1972. Diagr.

3210 _____. "Search for Geothermal Energy on Public Land May Stir Fight." New York Times, p. 20, Jan. 13, 1973.

3211 "Will Geothermal Steam Projects Sizzle or Fizzle?" Chemical Week 113(8): 57-59, Aug. 22, 1973. Photo.

3212 Wilson, Howard M. "California Geothermal Hunt Heats Up." Oil & Gas Journal 71(5): 70-72, Jan. 29, 1973. Map, photos., chart.

3213 Yoakum, Jim. "That There May Be R-O-O-M for All." Our Public Lands 21(4): 22-24, Fall 1971. Photo.

4

WIND ENERGY

4000 Allison, H. J. A Wind Energy Storage Technique Uti-
      lizing a Hydrogen Oxygen Electrolysis Cell System.
      Paper presented at Frontiers of Power Technology
      Conference, Oklahoma State University, Oct. 29,
      1968.

4001 "The Alternatives: R & D for New Energy Sources. "
      Economic Priorities Report 3(2): 23-26, May-June
      1972. Graph.

4002 Anderson, Richard J. "The Promise of Unconventional
      Energy Resources. " Battelle Research Outlook 4
      (1): 22ff (4pp.), 1972. Photos.

4003 Andiranov, V. M., and Bystritskii, D. N. "Parallel
      Operation of a Wind Power Station with a Powerful
      Grid. " Elektrichestvo (5): 8-12. Trans. by Elec-
      trical Research Association. No. IB 1161.

4004 _____, and Polataev, A. I. "Regulation of the Out-
      put of a Wind Power Station. " Elektrichestvo (6):
      19-24, 1952. Trans. by Electrical Research Asso-
      ciation, Trans, Trans. No. IB 1249.

4005 "Back to the Windmill to Generate Power. " Business
      Week (2330): 140-42, May 11, 1974. Photos.

4006 Barnhart, Earle. "Wind Power. " The Journal of the
      New Alchemists (1): 12-14, 1974. Plans, refs.

4007 Benson, Arnold. Plans for the Construction of a Small
      Wind Electric Plant. Stillwater, OK: Oklahoma
      State University Publications No. 33.

4008 Bergey, Karl H. "Wind Power Demonstration and
      Siting Problems. " Wind Energy Conversion System
      Workshop Proceedings, NSF/RA/W-73-006. pp.
      41-45, Dec. 1973.

148

4009 "Blowing in the Wind." Newsweek 77(25): 78-9, June
      21, 1971.

4010 Bodek, A. Performance Test of an 8-Meter-Diameter
      Andrean Windmill. Brace Research Institute, Pub.
      No. T. 12, 1964. 24pp.

4011 Bossell, Hartmut. Low Cost Windmill for Developing
      Nations. VITA, Oct. 1970

4012 Brand, David. "It's an Ill Wind, Etc.: Energy Crisis
      May be Good for Windmills." Wall Street Journal,
      p. 18, Jan. 11, 1974.

4013 Brangwyn, Frank. Windmills. Ann Arbor, MI: The
      Finch Press, 1923 Repr. $14.

4014 Burton, T. H. A Simple Electric Transmission Sys-
      tem for a Free Running Windmill. Brace Research
      Institute.

4015 Bush, Clinton G., Jr. "'Supercore' Eliminates House-
      hold Waste and Pollution at the Source." Catalyst
      for Environmental Quality 3(2): 22-26, 1973. Diagrs.

4016 Chilcoff, R. E. The Design, Development, and Test-
      ing of a Lowcost 10 hp Windmill Prime Mover
      Brace Research Institute, Pub. No. MT. 7, 1970.
      113pp.

4017 _____, et al. Specifications of the Brace 10-hp
      Airscrew Windmill, Brace Research Institute, Pub.
      No. T. 43, Feb. 1971. Plans, drawings.

4018 Clark, Wilson. "Interest in Wind is Picking Up as
      Fuels Dwindle." Smithsonian 4: 70ff (8pp.), Nov.
      1973. Photos.

4019 _____. "Return of the Windmill." Smithsonian,
      Nov. 1973.

4020 _____, et al. "How to Kick the Fossil Fuel
      Habit." Environmental Action 5(18): 3-6, 12-15,
      Feb. 2, 1974. Drawing.

4021 Clews, Henry M. Electric Power from the Wind.
      East Holden, MA: Solar Wind Publication, 1973.
      29pp.

4022 _____. "Solar Windmill. " Alternative Sources of
      Energy (8): Jan. 1973.

4023 _____. "Wind Power Systems for Individual Appli-
      cations. " Wind Energy Conversion Systems Work-
      shop Proceedings, NSF/RA/W-73-006, pp. 164-69,
      Dec. 1973.

4024 Coon, Dick. "Giant Wind Machine for Generating
      Electricity Gets Federal Scrutiny. " National Ob-
      server, p. 13, June 24, 1972.

4025 Design of High Speed Windmills Suitable for Driving
      Electrical Generators. National Research Labora-
      tories, Ottawa, Report No. PAA-32.

4026 DeVinck, Antoine. Wim of the Wind. Garden City,
      NY: Doubleday, 1974. Color illus. $5.95; PLB
      $.75 extra. (Gr. 1-3). 32pp.

4027 Duggal, J. S. Dynamic Analysis of High Speed Wind-
      Turbine Systems. Brace Research Institute, Pub.
      No. MT9, 1971. 12pp.

4028 Edmondson, William B. "A 100 KW Windmill Gen-
      erator. " Solar Energy Digest 2(4): 4, April 1974.

4029 "Energy Crisis: Are We Running Out?" Time 99(24):
      49-53, June 12, 1972. Graph, photos.

4030 Energy Primer: Solar, Water, Wind, and Bio Fuels.
      Berkeley, CA: Portola Institute, 1974. $4.50.
      (pap).

4031 "Energy Research and Development. " Report House
      Committee on Science and Astronautics, 92nd Con-
      gress II Serial EE, Dec. 1972. Diagrs., graphs,
      tables. 418pp.

4032 Fales, E. N. A New Propeller Type High Speed
      Windmill for Electrical Generation, ASME Trans-
      actions 19, Paper AER-50-6, 1928.

4033 Fateev, E. M. and Rozhdestvenskii, I. V. "Achieve-
      ments of Soviet Wind Power Engineering. " Vestnik
      Machinostzoeniya (9): 24-27, 1952. Electrical Re-
      search Association trans #IB 1334.

4034 Final Report on the Wind Turbine. Research Report
       PB 25370. New York University Office of Produc-
       tion Research and Development. Wah Production
       Board, Washington, DC, 1946.

4035 Finnegan, B. "Windmills." Hobbies 71: 118-19, Feb.
       1967. Illus.

4036 Ford, Kendall. "Wind Driven Generator." Popular
       Science Aug. 1938.

4037 Freese, Stanley. Windmills & Millwrighting. Cran-
       bury, NJ: Great Albion Books, 1972. Illus. $7.95.

4038 Fuller, R. Buckminster. "Energy Through Wind
       Power." New York Times, p. 39, Jan. 17, 1974.
       Photo.

4039 Ghaswala, S. K. "Windmill Design for Economic
       Power." Science News 93(6): 148, Feb. 10, 1968.

4040 Ghosh, P. K. Estimation of the Maximum Drag and
       Root Bending Moment on a Stationary Brace Air-
       screw Windmill Blade and Variation of Airscrew
       Starting Torque with Blade Angle. Brace Research
       Institute, Pub. No. CP. 19, May 1969. 13pp.

4041 _____. Low Drag Laminar Flow Aerofoil Section of
       Windmill Blades. Brace Research Institute, Pub.
       No. CP. 20, May 1969. 8pp.

4042 Gimpel, G., and Stodhart, A. H. Windmills for Elec-
       tricity Supply in Remote Areas. Electrical Re-
       search Association, Pub. No. C/T 120, 1958. 24pp.

4043 Glauert, H. "Windmills and Fans." Aerodynamic
       Theory, Vol. IV, W. F. Durand, ed. Dover Pub-
       lications, 1973. pp. 324-32.

4044 Golding, E. W. The Generation of Electricity by Wind
       Power. London: E. and F. N. Spon Ltd., 1955.
       318pp.

4045 _____, and Stodhart, A. H. The Potentialities of
       Wind Power for Electricity Generation. British
       Electrical and Allied Industries, Research Associa-
       tion, Technical Report No. W/T 16, 1949.

152 / Alternate Sources of Energy

4046 _____, and _____. The Use of Wind Power In
Denmark. Electrical Research Association, Pub.
No. C/T 112, 1954. 16pp.

4047 Hackleman, Michael. "The Savonius Super Rotor!"
Mother Earth News (26): 78-80, March 1974.
Photos., plans, chart.

4048 Hamilton, Bruce. "Hans Meyer and Windworks."
High County News 6(13): 15-16, June 21, 1974.
Photos.

4049 Helical Sail Windmill. VITA, pub. no. 11131.1.

4050 Heronemus, William E. "Alternatives to Nuclear Energy." Catalyst for Environmental Quality 2(3):
21-15, 1972. Drawing, photo.

4051 Hewson, E. Wendell. First Progress Report on Research on Wind Power Potential in Selected Areas
of Oregon. Oregon State University, Report No.
PUD 73-1, March 1973.

4052 Hicks, Nancy. "Energy Crisis Impels Many to Study
and Erect Windmills as Power Source." New York
Times, p. 33, May 20, 1974.

4053 Holt, H. H. "Windmill Island: Windmill Island City
Park, Holland, Michigan." American Forests 73:
24-26+, Dec. 1967. Illus.

4054 Hubbert, M. King. "The Energy Resources of the
Earth." Scientific American 224(3): 60ff (11pp.),
Sept. 1971. Charts, diagrs., photo.

4055 Hütter, Ulrich. The Aerodynamic Layout of Wind
Blades of Wind Blades of Wind Turbines with High
Tip Speed Ratios. United Nations Conference on
New Sources of Energy, Rome, 1961. New York:
United Nations, 1964.

4056 _____. "The Development of Wind Power Installations for Electrical Power Generation in Germany."
Brennstoff-Waerme-Kraft 6(7): 270-78, 1954. Trans.
by Scientific Translation Corp., Santa Barbara, CA.
NTIS #N73-29009/OGA.

4057 _____ . "Operating Experience Obtained with a 100
KW Wind Power Plant. " Brennstoff-Waerme-Kraft
16: 333-40, 1964. Trans. by Kanner Associates. NTIS #N73-29008/2GA.

4058 _____ . "Past Developments of Large Wind Generators in Europe. " Wind Energy Conversion Systems Workshop Proceedings, NSF/RA/W-73-006,
pp. 19-22, Dec. 1973.

4059 _____ . "Practical Experience Gained from the De-
velopment of a 100-KW Wind Power Installation. "
Brennstoff-Waerme-Kraft 16(7): 333-341, July 1964.
Trans. by G. T. Ward, Brace Research Institute,
Pub. No. T. 42.

4060 Iwasaki, Matsunosuke. The Experimental and Theoretical Investical Investigation of Windmills. Report of Research Institute for Applied Mechanics,
No. 8. Japan, 1953. pp. 181-229.

4061 Kadivar, M. S. Potential for Wind Power Development. Brace Research Institute, Pub. No. MT.
10, July 1970. 171pp.

4062 Kidd, Stephen and Garr, Doub. "Can We Harness
Pollution-Free Electric Power from Windmills?"
Popular Science 201(5): 70-73, Nov. 1972. Draw-
ings, photos.

4063 Kloeffler, R. G. and Stiz, E. L. "Electric Energy
from the Wind. " Kansas State College Bulletin
(9): Sept. 1946.

4064 Kogan, A., and Nissim, E. Shrouded Aerogenerator
Design Study 1, Haifa Institute of Technology, Dept.
of Aeronautical Engineering, Report 17, Jan. -June
1961.

4065 Krasovskiy. "Project of Wind Motor with Aerodynamic
Transmission ofr Capacities of 100KW to 3000KW."
Izvestia Otd. Tekh. Nauk, Akad. Nauk SSR (USSR)
(5): 65-77, 1939. Trans. by Kanner Associates,
Redwood City, CA. NTIS #N73-30976/7GA.

4066 Kroms, A. "Wind Power Stations Working in Connec-
tion with Existing Power Systems. " Alternative

Sources of Energy Bulletin 45(5): 135-144, March 1954. Electrical Research Association Trans. No. IB 1371.

4067 Kushul, Mikhail Y. Self-Induced Oscillation of Rotors. New York: Plenum, 1964. $25.

4068 Levy, N. Current State of Wind Power Research in the Soviet Union. Brace Research Institute, Pub. No. T. 56, Sept. 1968.

4069 "Lighting Some Candles. " Architectural Forum 139(1): 89ff (10pp.), July-Aug. 1973. Drawing; photos.

4070 Lilley, G. M. , and Rainbird, W. J. A Preliminary Report on the Design and Performance of Ducted Windmills. British Electrical and Allied Industries Research Association, Technical Report C/T 119, 1957. 65pp.

4071 Lindsley, E. F. "Quirk: High Output Wind Generation. " Popular Science 183: 847-49, March 1, 1974. Illus.

4072 _____. "Wind Power: How New Technology is Harnessing an Age-Old Energy Source. " Popular Science 205(1): 54ff (5pp.), July 1974. Diagrs. , photos.

4073 McCaull, Julian. "Windmills. " Environment 16(1): 6ff (12pp.), Jan. -Feb. 1973. Drawing, photos.

4074 McColly, H. F. , and Buck, Foster. "Homemade Six-Volt Wind-Electric Plants. North Dakota State University, 1936. " Available from the Mother Earth News, Hendersonville, NC.

4075 Mattioli, G. D. Theory and Calculation of a Wind Turbine. Trans. by Electrical Research Association. Trans. No. IB 1381.

4076 May, R. O. "Origins of Feedback Control; Mechanisms for Controlling Windmills. " Scientific American 223: 115-18, Oct. 1970. Illus.

4077 Merrill, Richard, et al. Renewable Sources: and Energy Primer. Berkeley, CA: Portola Institute, 1974.

Illus. $4. 112pp.

4078 Meyer, Hans. Wind Energy. Domebook 2, Pacific Domes, Box 279, Bolinas, CA 94924.

4079 _____. "Wind Energy." Architectural Design 42: 773, Dec. 1973. Illus.

4080 _____. "Wind Generators: Here's an Advanced Design You Can Build." Popular Science 201: 103-05+, Nov. 1972. Illus.

4081 Morrison, J. G. The Development of a Method for Measurement of Strains on the Blades of a Windmill Rotor. British Electrical and Allied Industries Research Association. Technical Report C/T 117, 1957. 28pp.

4082 Nakra, H. L. A Report on Preliminary Testing of a Lubing Windmill Generator (M022-3G024-400) of the Brace Research Institute. Brace Research Institute, Pub. No. T. 75, 5pp.

4083 Noel, John M. "French Wind Generator Systems." Wind Energy Conversion System Workshop Proceedings, NSF/RA/W-73-006, pp. 186-96, Dec. 1973.

4084 Pickering, William H. "Important to Examine Possibilities of Unconventional Energy Sources." Aware 31(5): 5-7, April 1873.

4085 Plumlee, R. H. Perspectives in U. S. Energy Resource Development. Albuquerque, NM: Sandia Labs, 1973. 113pp. NTIS #SLA-73-163/GA.

4086 Preston, D. J. "One Man's Answer to the Energy Crisis: Spocott Windmill." American Forests 80 (2): 20-23, 65-67, Feb. 1974. Illus.

4087 Proceedings of NASA's Wind Energy Program, NASA, Lewis Research Center, Mail Stop 500-201, Cleveland, OH 44135

4088 Putnam, Palmer Cosslett. Power from the Wind. New York: Van Nostrand Reinhold, 1974. 240pp. $9-95.

4089 Quigg, P. W. "Alternative Energy Sources." Satur-
      day Review World 1(12): 32, Feb. 23, 1974.

4090 Ramatahan, R. "The Economics of Generating Elec-
      tricity from the Wind Power in India." Indian
      Quarterly Journal of Economics, pp. 245-55, Jan.
      1970.

4091 Reynolds, John. Windmills & Watermills. New York:
      Praeger Publishers, 1970. Illus. 196pp. $13.95.

4092 Roddis, Louis H., Jr. "Unconventional Power
      Sources." Survey Report Presented at Edison
      Electric Institute Executives Conference, Scotts-
      dale, AZ, Dec. 4, 1970.

4093 Sanches-Vilar, C. Wind Electric Research Report
      (Evaluation of the Performance of a Commercial
      Aero-generator). Brace Research Institute, Pub.
      No. DT. 4.

4094 Savonius, S. J. "The S-Rotor and Its Applications."
      Mechanical Engineering 53(5): 333-338, May 1931.

4095 Savonius Rotor Windmill. VITA, Pub. No. 11132.1.

4096 Schumann, W. A. "Wind Generator Study Funds
      Sought." Aviation Week and Space Technology 102
      (2): 57-58, Jan. 13, 1975. Photo.

4097 Schwartz, Mark. "Can Windmills Supply Farm
      Power?" Organic Gardening and Farming 21(1):
      155-58, Jan. 1974. Photo.

4098 Sektorov, V. R. "The Present State of Planning and
      Erection of Large Experimental Wind Power Sta-
      tions." Elektrichestvo (2): 9-13, 1933. Trans. by
      Electrical Research Association, Trans. No. IB
      1052.

4099 Sencenbaugh, Jim. "I Built a Wind Charger for $400!"
      Mother Earth News (20): 32-36, March 1973. Photo,
      plans.

4100 Shefter, Ya., et al. "Solar and Wind Power to be
      Harnessed." (Energiyu Solntsa i Vetra-V
      UPRYAZHU) Pravda, Nov. 1, 1972. 5pp. Trans.

by Army Foreign Science and Technology Center, Charlottesville, VA. NTIS #AD-765 783/6.

4101 "Shell Platforms Utilize Wind and Sun Power." Petroleum Engineering International 46: 14, Sept. 1974.

4102 "Shell Tests Windmill Generators Offshore." Oil & Gas Journal 72(36): 92, Sept. 9, 1974. Photos.

4103 Sherman, Marcus. "A Windmill in India." The Journal of the New Alchemists (1): 15-17, 1974. Drawings.

4104 Shrouded Aerogenerator Design Study 2, Israel Society of Aeronautical Sciences, 5th Conference on Aviation and Astronautics, 1963. p. 49-56.

4105 Shumann, W. A. "NASA Spurs Wind Generator Program." Aviation Week and Space Technology 100 (13): 41, April 1, 1974.

4106 Simonds, M. A 1-KW Wind Driven Domestic Lighting Plant at Spring Head, St. James, Barbados. Brace Research Institute. Pub. No. T. 7, 1964. 11pp.

4107 _____, and Bodek, A. Performance Test of a Savonius Rotor. Brace Research Institute, Pub. No. T. 10, Jan. 1964. 20pp.

4108 Smith, B. E. "Smith-Putnam Wind Turbine Experiment." Wind Energy Conversion Systems Workshop Proceedings, NSF RA/W-73-006, pp. 5-7, Dec. 1973.

4109 Smith, Gerry, ed. List of Past and Present Windmill Manufacturers. Cambridge, Great Britain: University of Cambridge, 1 Scroope Terrace, Cambridge CB2 1 PX.

4110 South, P., and Rangi, R. S. Preliminary Tests of a High Speed Vertical Access Windmill Model. National Research Council of Canada, Laboratory Technical Report LTR-LA-74, March 1971.

4111 Spanides, A. G., and Hatzkakidis, A. D., eds. International Seminar on Solar and Aeolian Energy, 1961. New York: Plenum, 1964. 491pp. (Pro-

ceedings of Advanced Study Institute for Solar and
Aeolian Energy, Sounion, Greece.)

4112 Spier, Peter. Of Dikes and Windmills. Garden City,
NY: Doubleday, 1969. Illus. (Gr. 6-9) $5.95.

4113 Stabb, D. "Wind." Architectural Design 43(42): 253-
54, April 1972. Illus., diagrs.

4114 Sterne, L. H. G., and Rose, G. C. The Aerodynamics
of Windmills Used for the Generation of Electricity.
Electrical Research Association, Pub. No. IB/T,
1951. 12pp.

4115 Stoner, Carol Hupping, ed. Producing Your Own
Power: How to Make Nature's Energy Sources
Work for You. Rodale Press Books (Organic Gardening & Farming), Sept. 1974. 229pp. Illus.,
refs., index. $8.95.

4116 Swanson, Robert K., et al. Wind Power Development
in the United States. N.p., n.d.

4117 Tagg, J. R. Wind-Driven Generators: The Difference
Between the Estimated Output and Actual Energy
Obtained. Electrical Research Association, Pub.
No. C/T 123, 1960.

4118 "Thermal Energy Creates Rotational Motion." Machine
Design 46(23): 40, Sept. 19, 1974. Diagr.

4119 Thomas, Percy H. "Electric Power from the Wind."
Washington, DC: Federal Power Commission,
March 1945.

4120 Thomas, Ronald L. "Wind Energy Conversion."
Energy Environment Productivity. Proceedings of
the First Symposium on Rann: Research Applied to
National Needs, Washington, DC, Nov. 18-20, 1973.
Washington, DC: National Science Foundation, pp.
42-45. Diagr., drawings, map, charts, graphs.

4121 Todd, John H. "A Modest Proposal." The New Alchemists. 20pp. Drawings, graphs, refs. Avail.
from The New Alchemy Institute-East, Box 432,
Woods Hole, MA 02543.

4122 Tompkin, J. "Introduction to Voigt's Wind Power Plant." Wind Energy Conversion Systems Workshop Proceedings, NSF/RA/W-73-006, pp. 23-26, Dec. 1973.

4123 Tondl, Alex. Some Problems of Rotor Dynamics. New York: Barnes & Noble, Inc., 1965. $16.

4124 Trunk, E. "Free Power from the Wind: A Short Survey of Windmills." The Mother Earth News 60-64, Sept. 1972.

4125 Two Plans for a Fan-Bladed Turbine Type Windmill with Applications. VITA.

4126 Udall, S. L. "Homage to Windmills." Reader's Digest 98: 95-96, May 1971. Illus.

4127 United Nations. New Sources of Energy and Economic Development: Solar Energy, Wind Energy, Tidal Energy, Geothermal Energy, and Thermal Energy of the Seas. New York: United Nations, 1957.

4128 _____. New Sources of Energy and Energy Development. Report on the United Nations Conference on New Sources of Energy: Solar Energy--Wind Power--Geothermal Energy; Rome, 21-31 Aug. 1961. 65pp. New York: United Nations, 1962. $.75.

4129 _____. Proceedings of the United Nations Conference on New Sources of Energy: Solar Energy, Wind Power, and Geothermal Energy, Rome, 21-31, Aug. 1961. New York: United Nations, 1964. Vol. VII: Wind Power. Illus. 408pp.

4130 _____. Educational, Scientific and Cultural Organization (UNESCO). Wind and Solar Energy Proceedings of the New Delhi Symposium. Paris: UNESCO, 1956. Illus. 128pp.

4131 US/AID Fan Blade Windmill. VITA, Pub. No. 11133.3

4132 Villers, D. E. The Testing of Wind Driven Generators Operating in Parallel with a Network. Electrical Research Association, Pub. No. C/T 116, 1957. 22p.

4133   Vince, John. Windmills. New York: International
       Publications Service, 1969. $1 (pap).

4134   Vita-Essex Windmill. VITA, Pub. No. 11133.1.

4135   Voigt, H. "Principles of Steel Construction in the
       Building and Operation of Wind Driven Power
       Plants." Der Stahlbau 23(8): 184-88, Aug. 1954.
       Trans. by Kanner Assoc., Redwood City, CA.
       NTIS #N73-21253.

4136   Wade, Nicholas. "Windmills: The Resurrection of an
       Ancient Energy Technology." Science 184(4141):
       1055-58, June 7, 1974, Diagrs.

4137   Wailes, Rex. English Windmill. Clifton, NJ: Augus-
       tus M. Kelley, publisher, 1967. Illus. $12.50.

4138   Walcott, John. "William Heronemus, Unlikely Revolu-
       tionary." Blair & Ketchum's Country Journal 1(7):
       22-27, Nov. 1974. Photos., diagrs.

4139   Wilcox, Carl. "Motion Pictures History of the Erec-
       tion and Operation of the Smith-Putnam Wind Gen-
       erator." Wind Energy Conversion Workshop, NSF/
       RA/W-73-006. Savino, J. M., ed. pp. 8-10,
       Dec. 1973.

4140   Wind Bibliography. Alternative Sources of Energy,
       No. 9.

4141   "Wind Energy Studies Started by NASA." Aviation
       Week and Space Technology 101(23): 49, Dec. 9,
       1974.

4142   "Wind Generators Use Sun's Energy More Effectively
       Than Solar Cells." Electrical Review 192(22): 782-
       84, June 1, 1973.

4143   World Meteorological Organization. Sites of Wind-
       Power Installations. New York: Unipub, 1964.
       $2.50 (pap).

5

TIDAL ENERGY

5000 Anderson, Richard J. "The Promise of Unconventional
Energy Resources. " Battelle Research Outlook
4(1): 22ff (4pp.), 1972. Photos.

5001 Barber, N. F. Water Waves. London and Winchester,
England: Wykeham Publications, 1969. (pap) (Juv.)

5002 Bernstein, Lev. "Energy of the Northern Seas. "
Marine Technology Society Journal 7: 24, Oct. -
Nov. 1973.

5003 Clancy, Edward P. The Tides. Garden City, NY:
Doubleday, 1968. Photos., drawings, charts,
diagrs., index. 228pp.

5004 Clark, Wilson. Energy for Survival: Alternative to
Extinction. Garden City, NY: Anchor Press/
Doubleday. Illus. $12.50.

5005 _____, et al. "How to Kick the Fossil Fuel Habit."
Environmental Action 5(18): 3-6, 12-15, Feb. 2,
1974. Drawing.

5006 Darwin, George H. The Tides. San Francisco:
W. H. Freeman and Co., 1962. 369pp.

5007 Davey, N. Studies in Tidal Power. London, England:
Constable, 1923. 255pp.

5008 Defant, Albert. Physical Oceanography, Vol. II. New
York: Pergamon Press, 1961. 570pp.

5009 Doodson, A. T., and Warburg, H. D. Admiralty
Manual of Tides. London, England.

5010 Duxbury, A. C. The Earth and Its Ocean. Reading,
MA: Addison-Wesley Publishing Co., Inc. 1971.

161

✳ 5011 "Energy Crisis: Are We Running Out?" Time 99(24): 49-53, June 12, 1972. Graph, photos.

5012 Energy Primer: Solar, Water, Wind & Bio Fuels. Berkeley, CA: Portola Institute, 1974. $4.50 (pap.)

5013 Evans, D. G., and McCutchan, J. C. "Electricity Generation in Australia." Search 2(9): 338ff (8pp.), Sept. 1971. Maps, charts, graphs.

5014 Gannon, Robert. "Atomic Power, What Are Our Alternatives?" Science Digest 70(6): 18ff (5pp.), Dec. 1971. Photos.

5015 Gray, T. J. and Gashus, O. K. International Conference on the Utilization of Tidal Power. New York: Plenum, 1972. 630pp.

5016 Greager, W. P., and Justin, J. D. Hydroelectric Handbook. New York: Wiley, 1950.

5017 Haswell, Charles K., et al. "Pumped Storage and Tidal Power in Energy Systems." Journal of the Power Division (ASCE) 98: 201ff (20pp), Oct. 1972. Map, diagrs., graphs, refs.

5018 Hidy, G. M. The Waves. New York: Van Nostrand Reinhold, 1971.

5019 Hubbert, M. King. "Energy Resources: A Report to the Committee on Natural Resources of the National Academy of Sciences--National Research Council, Washington, DC: NAS--National Research Council, 1962. 153pp.

✳ 5020 _____. "The Energy Resources of the Earth." Scientific American 224(3): 60ff (11pp.), Sept. 1971. Charts, diagrs., photo.

5021 Hübner, R. "Tidal Power Stations." Elektrotechnische Zeitschrift 24(1): 485ff (5pp.), Sept. 5, 1972. Diagrs., drawings, graph, photos., table.

5022 "Interest Renewed in Fundy Tiday Power Scheme." Engineering News-Record 186(5): 15, Feb. 4, 1971.

5023 Macmillan, D. H. Tides. New York: Elsevier, 1966.

Graphs, diagrs., photos, index. 240pp.

5024 Nash, Barbara. "Energy from the Earth and Beyond." Chemistry 46(9) 6ff (4pp.), Oct. 1973. Diagrs., map, photo.

5025 Neumann G. Ocean Currents. New York: Elsevier, 1968. Refs.

↯ 5026 Paton, T. A. L., and Brown, J. G. Power from Water. London, England: L. Hill, 1961.

5027 Quigg, P. W. "Alternative Energy Sources." Saturday Review World 1(12): 32, Feb. 23, 1974.

5028 Russell, R. C. H., and Macmillan, D. H. Waves and Tides. London, England: Hutchinson Scientific and Technical Publications, 1954.

5029 Shaw, T. L., and Thorpe, G. R. "Integration of Pumped-Storage with Tidal Power." Proceedings of the American Society of Civil Engineers 97P01 (7810): 159-80, Jan. 1971. Maps, diagrs. Discussion by M. H. Chaudhry, Proceedings of the American Society of Civil Engineers 98P01(8927): 155-66, June 1972.

5030 Smil, Vaclav. "Energy and the Environment." Futurist 8(1): 4-13, Feb. 1974. Photos, diagrs.

5031 Smith, F. G. Walton. The Seas in Motion. New York: Thomas Y. Crowell, Co., 1973. Photos, diagrs., charts, index. 248p.

5032 Sverdrup, H. U., Johnson, M. W., and Fleming, R. H. The Oceans. Englewood Cliffs, NJ: Prentice-Hall, 1942.

5033 "Tidemills in England and Wales." Transactions of the Newcome Society 16: 1-33, 1935-1936.

5034 United Nations. New Sources of Energy and Economic Development: Solar Energy, Wind Energy, Tidal Energy, Geothermal Energy, and Thermal Energy of the Seas. New York: United Nations, 1957. Refs.

5035 Wailes, R. <u>Tidemills</u>. London, England: Society for
the Protection of Ancient Buildings, 1957.

✳ 5036 Weaver, Kenneth F. "The Search for Tomorrow's
Power." <u>National Geographic</u> 142(5): 650ff (32pp.),
Nov. 1972. Photos, diagrs.

6

ENVIRONMENTAL ARCHITECTURE
(Climate and Energy)

6000 Abrahamson, D. E., and Emmings, S., eds. Energy
      Conservation: Implications for Building Design and
      Operation. Minneapolis, MN: Minneapolis Univer-
      sity of Minnesota, School Public Affairs, 1973.
      156pp.

6001 Abrams, C. Housing in the Modern World. London:
      Faber, 1966. 307pp.

6002 _____. Man's Struggle for Shelter in an Urbanizing
      World. Cambridge, MA: Massachusetts Institute
      of Technology, 1964. 307pp.

6003 Aronin, J. E. Climate and Architecture. New York:
      Reinhold, 1953. 304pp.

6004 Banham, R. Architecture of the Well-tempered En-
      vironment. Chicago: University of Chicago, 1969.
      295pp.

6005 Chahroudi, Day, and Wellesley-Millis, Sean. Climatic
      Envelopes. Paper presented to the American In-
      stitute of Architects, National Convention, 1974.
      Available from MIT, Dept. of Architecture. Paper
      dated May 23, 1973.

6006 Climatology and Building. Vienna: Proceedings of the
      CIB Symposium, 1970/71. 279pp.

6007 Doxiadis, C. A. "Ecumenopolis, World-City of To-
      morrow." The Ecology of Man: An Ecosystem
      Approach, pp. 154-62. Smith, L., ed. New York:
      Harper & Row, 1972.

6008 Fry, E. M., and Drew, J. Tropical Architecture in
      the Dry and Humid Zones. New York: Reinhold,
      1964. 264pp.

6009 _____, and _____. Tropical Architecture in the
      Humid Zone. New York: Reinhold, 1956. 320pp.

6010 Giedion, S. The Beginnings of Architecture. New
      York: Pantheon, 1964.

6011 Givoni, B. Man, Climate and Architecture. New
      York: Elsevier, 1969. 364pp.

6012 Gottlieb, L. D. Environment and Design in Housing.
      New York: Macmillan, 1965. 258pp.

6013 Harrison, J. D. Environmental Preferences: Relevant
      Studies for Urban Planning. (C. P. L. Bibliography),
      1973. 19pp.

6014 Holloway, Dennis S., et al. OUROBOROS East:
      Towards an Energy Conserving Urban Dwel-
      ling. Minneapolis: University of Minnesota, School
      of Architecture and Landscape Architecture, 1974.
      $5.

6015 Huntington, E. Civilization and Climate. New Haven:
      Yale University, 1924.

6016 Manners, G. Geography of Energy. London: Hutch-
      inson, 1964. Chicago: Aldine, 1967. 205pp.
      London: Hutchinson, 1971. 222pp.

6017 Markham, S. F. Climate and the Energy of Nations.
      New York: Oxford University Press, 1947. 240pp.

6018 Miller, W. C. Factors and Forces Influence to Arch-
      itectural Design (C. P. L. Bibliography) No. 124.

6019 National Research Council. Weather and the Building
      Industry. Washington, DC: National Research
      Council, Building Research Advisory Board, 1950.
      158pp.

6020 Olgyay, V. Design with Climate; bioclimatic approach.
      Princeton, NJ: Princeton University, 1963. 190pp.

6021 Oliver, P. Shelter and Society. Barrie and Rickliffe,
      Cresset Press, 1969.

6022 Predicasts, Inc. Energy Savings in Buildings. Cleveland, OH: Predicasts, Inc., 1974. $375.00.

6023 Rapoport, A. "The Ecology of Housing." Ecologist
      3(1): 10-17, 1973.

6024 _____. House Form and Culture. Englewood
      Cliffs, NJ: Prentice-Hall, 1969. 150pp.

6025 United Nations. Department of Economic and Social
      Affairs. Climate and House Design. New York:
      United Nations, 1971. 93pp.

6026 VanStraaten, J. F. Thermal Performance of Build-
      ings. New York: Elsevier, 1967. 311pp.

6027 Wilson, R., and Jones, W. J. Energy, Ecology and
      the Environment. New York: Academic Press,
      1974. 300pp.

NOTE: For other works on Environmental architecture, see
       following entries in Section 2, "Solar Energy":

1003	1014	1018	1019	1118
1120	1182	1230	1320	1431
1492	1543	1634	1666	1677
1693	1698	1717	1830	1832
1838	1889	1894	1914	1941
1942	1958	1979	1978	2008
2011	2031	2084	2086	2087
2088	2089	2090	2092	2093
2149	2150	2151	2157	2174
2176	2209	2218		

Appendix

ADDRESSES OF RELEVANT PERIODICALS
ORGANIZATIONS, AND INDIVIDUALS

AEG TELEFUNKEN PROGRESS
Elitera-Verlag
Fritz-Wildung Str. 22
1 Berlin 33, W. Germany

AERONAUTICAL JOURNAL
Royal Aeronautical Society
4 Hamilton Place
London W1V OBQ, England

AGRICULTURE ENGINEERING
American Society for Engineer-
ing Education
Dupont Circle Bldg.
1346 Connecticut Avenue, N.W.
Washington, DC 20036

AIR CONDITIONING, HEATING,
AND REFRIGERATION NEWS
Business News Pub. Co.
P.O. Box 6000
Birmingham, MI 48012

AIChE WATER SYMPOSIUM
SERIES
American Society of Chemical
Engineers
345 East 47th Street
New York, NY 10017

ALTERNATE SOURCES OF
ENERGY
c/o Donald Marier
Route 1, Box 36B
Minong, WI 54859

ALTERNATIVES
c/o Trent University
Peterborough
Ontario, Canada

AMBIO
Royal Swedish Academy of
Science
Universitetsforlaget, Box 307
Blindern
Oslo 3, Norway

AMERICAN ACADEMY OF ARTS
& SCIENCES
Records
280 Newton Street
Brookline Station, MA 02146

AMERICAN FORESTS
American Forestry Assn.
1319 18th Street, N.W.
Washington, DC 20036

AMERICAN GAS ASSOCIATION
MONTHLY
American Gas Association
1515 Wilson Blvd.
Arlington, VA 22209

AMERICAN GEOPHYSICAL
UNION TRANSACTIONS
American Geophysical Union
1717 L Street, N.W.
Washington, DC 20036

AMERICAN INSTITUTE OF
ARCHITECTS
Solar Energy Program
1735 New York Avenue, N.W.
Washington, DC 20006

AMERICAN INSTITUTE OF AS-
TRONAUTICS & AERONAUTICS
1290 Avenue of the Americas
New York, NY 10019

AMERICAN JOURNAL OF
  SCIENCE
Kline Geology Laboratory
Yale University
New Haven, CT   06520

AMERICAN MINING CONGRESS
  see MINING CONGRESS
  JOURNAL

AMERICAN PUBLIC POWER
  ASSOCIATION
Suite 212
2600 Virginia Avenue, N.W.
Washington, DC   20037

AMERICAN SCIENTIST
Society of the Sigma XI
155 Whitney Avenue
New Haven, CT   06510

AMERICAN SOCIETY OF CIVIL
  ENGINEERS
345 E. 47th Street
New York, NY   10017

AMERICAN SOCIETY OF
  MECHANICAL ENGINEERS see
  ASME

THE AMERICAN WAY
American Hospital and Life In-
  surance Co.
P.O. Box 2341
San Antonio, TX   78206

ANNALS OF THE AMERICAN
  ACADEMY OF POLITICAL AND
  SOCIAL SCIENTISTS
3937 Chestnut Street
Philadelphia, PA

APPLIED OPTICS
Optical Society of America
355 East 46th Street
New York, NY   10017

APPLIED SOLAR ENERGY
Allerton Press
150 5th Avenue
New York, NY   10011

ARCHITECTURAL DESIGN
Standard Catalogue Co., Ltd.
26 Bloomsbury Way
London WC1A, England

ARCHITECTURAL FORUM
Whitney Publications
130 East 59th Street
New York, NY   10022

ARCHITECTURAL RECORD
McGraw-Hill
1220 Avenue of the Americas
New York, NY   10020

l'ARCHITECTURE d'AUJOURD-
  'HUI
5 Rue Bartholdi
Boulogne (Seine)
France

ARCHITECTURE PLUS
Informat Publishing Corporation
1345 Avenue of the Americas
New York, NY   10019

ASCE see AMERICAN SOCIETY
  OF CIVIL ENGINEERS

ASHRAE ENGINEERS REPORT
American Society of Heating,
  Refrigeration, and Air Condi-
  tioning
345 East 47th Street
New York, NY   10017

ASHRAE JOURNAL
(see above)

ASME TRANSACTIONS
American Society of Mechanical
  Engineers
345 East 47th Street
New York, NY   10017

ASTRONAUTICS AND AERO-
  NAUTICS see AMERICAN
  INSTITUTE OF ASTRONAUTICS
  AND AERONAUTICS

ATOM (U. S.)
Los Alamos Scientific Lab,
  University of California
Box 1663
Los Alamos, NM   87544

AVIATION WEEK AND SPACE
  TECHNOLOGY
McGraw-Hill
1221 Avenue of the Americas
New York, NY   10020

AWARE
Box 24904
Los Angeles, CA   90024

BELL TELEPHONE LABORA-
  TORIES
Murray Hill, NJ   07971

BRACE RESEARCH INSTITUTE
MacDonald College of McGill
  University
Ste. Anne de Bellevue   800
Quebec, Canada

BRENNSTOFF-WÄRME-KRAFT
Zeitschrift für Energietechnik
  und Energiewirtschaft
DM. 114. VDI Verlag GmbH
Postfach 1129
4-Duesseldorf 1, West Germany

BUILDING SYSTEMS DESIGN
Industrial Press
235 Duffield St.
Brooklyn, NY   11201

BULLETIN OF THE ATOMIC
  SCIENTISTS
Education Foundation for Nuclear
  Scientists
935 E. 60th Street
Chicago, Ill   60637

BUSINESS WEEK
McGraw-Hill
1221 Avenue of the Americas
New York, NY   10020

CANADIAN MINING AND
  METALLURGY BULLETIN
Canadian Institute of Mining and
  Metallurgy
906-1117 Ste. Catherine St. W.
Montreal, 110, Quebec, Canada

CASABELLA
G. Milani s. a. s. Editrice
Via Marconi   17
20090 Segrate
Milan, Italy

CATALYST FOR ENVIRON-
  MENTAL QUALITY
274 Madison Avenue
New York, NY   10016

CHAROUDI, DAY
solar energy engineer
Zome Works Corp.
Albuquerque, NM

CHEMICAL AND ENGINEERING
  NEWS
American Chemical Society
1155 16th Street, N.W.
Washington, DC 20036

CHEMICAL ENGINEERING
McGraw-Hill
1221 Avenue of the Americas
New York, NY 10020

CHEMICAL ENGINEERING
  PROGRESS
American Institute of Chemical
  Engineers
345 East 47th Street
New York, NY   10017

CHEMICAL TECHNOLOGY
American Chemical Society
1155 16th Street, N.W.
Washington, DC   20036

CHEMICAL WEEK
McGraw-Hill
1221 Avenue of the Americas
New York, NY   10020

CHEMISTRY
Pergamon Press
44-01 21st Street
Long Island City, NY 11101

CHEMTECH
American Chemical Society
1155 16th Street, N.W.
Washington, DC 20036

CHRISTIAN SCIENCE MONITOR
Christian Science Publishing
Society
1 Norway Street
Boston, MA 02115

CIVIL ENGINEERING MAGAZINE
American Society of Civil
Engineers
345 East 47th Street
New York, NY 10468

CLEVITE CORPORATION
540 East 105th Street
Cleveland, OH 44108

COAL AGE
McGraw-Hill
1221 Avenue of the Americas
New York, NY 10020

COMBUSTION
277 Park Avenue
New York, NY 10017

COMMERCE TODAY
Supt. of Documents
Washington, DC 20402

COMPRESSED AIR MAGAZINE
942 Memorial Parkway
Phillipsburg, NJ 08865

CONGRESSIONAL QUARTERLY
1735 K Street, N.W.
Washington, DC 20006

CONSERVATION FOUNDATION
LETTER
1717 Massachusetts Ave., N.W.
Washington, DC 20036

CONSERVATIONIST
N.Y. State Dept. of Environmental Conservation
50 Wolf Road
Albany, NY 12201

CONSERVATION NEWS
National Wildlife Federation
1412 16th Street, N.W.
Washington, DC 20036

CONSULTING ENGINEER
R. W. Roe and Partners
217 Wayne Street
St. Joseph, MI 49085

CRY CALIFORNIA
California Tomorrow
Monadnock Bldg.
681 Market Street, Rm 1059
San Francisco, CA 94105

CURRENT
Plainfield, VT 05667

DOMUS
Editoriale Domus
Via Monte di Pieta 15
20121 Milan, Italy

DRILLING CONTRACTOR
International Association of
Drilling Contractors
5th Floor
211 N. Ervay Bldg.
Dallas, TX 75201

ECOLOGIST
73 Molesworth St.
Wadebridge, Cornwall
England

ECOLOGY TODAY
Ecological Dimensions, Inc.
Box 180
West Mystic, CT 06388

ECONOMIC PRIORITIES REPORT
84 Fifth Avenue
New York, NY 10011

THE ECONOMIST
Economist Newspaper Ltd.
25 St. James St.
London SW1A 1HG
England

EDMUND SCIENTIFIC CORP.
Barrington, NJ
(Fresnel lenses, solar-furnace
  kits, survival-still kits,
  aluminum foil, solar batteries,
  electric motors)

EDUCATION MATERIALS &
  EQUIPMENT CO.
Box 63
Bronxville, NY
(solar-demonstration kit)

EKISTICS
Athens Center Ekistics
Box 471
Athens, Greece

ELECTRICAL RESEARCH
  ASSOCIATION
Cleeve Road
Leatherhead Surrey
England

ELECTRICAL REVIEW
Thomas Skinner & Co., Ltd.
111 Broadway
New York, NY 10006

ELECTRICAL WORLD
McGraw-Hill
1221 Avenue of the Americas
New York, NY 10020

ELECTROCHEMICAL SOCIETY
  BULLETIN; also their
  JOURNAL
30 East 42nd Street
New York, NY 10017

ELECTRONIC ENGINEER
Chilton Co.
Chestnut and 56th Streets
Philadelphia, PA

ELECTRONICS & POWER
PPL
Box 8, Southgate House
Stevenage S91 1HG
Herts, England

ELEKTROTECHNISCHE
  ZEITSCHRIFT
Verband Deutscher Elektro-
  techniker
VDE-Verlag GmbH
Bismarck Str. 33
Berlin-Charlottenburg
West Germany

ELECTRONICS
McGraw-Hill
1221 Avenue of the Americas
New York, NY 10020

ELEKTRICHESTVO
B. Cherkassi per 2/10
Moscow, U.S.S.R.

ELECTRONICS LETTERS
Institution of Electrical
  Engineers
P.O. Box 8
Southgate House
Tevenage, Herts, England

ENERGY DIGEST
Scope Publications
1120 National Press Building
Washington, DC

ELGIN WATCH COMPANY
Elgin, IL
(Solar Clock)

ENGINEER
Georgia Technical Institute
Atlanta, GA 30332

ENGINEERING
Canadian Technical Publications
  Ltd.
17 Inkerman Street
Toronto 5, Canada

ENGINEERING AND MINING
 JOURNAL
McGraw-Hill
1221 Avenue of the Americas
New York, NY 10020

ENGINEERING JOURNAL
University of Massachusetts
Amherst, MA 01002

ENGINEERING NEWS-RECORD
McGraw-Hill
1221 Avenue of the Americas
New York, NY 10020

ENVIRONMENT
Committee for Environmental
 Information
438 N. Skinker
St. Louis, MO 63130

ENVIRONMENTAL QUALITY
 MAGAZINE
(Environmental Awareness Associates, Woodland Hills, CA;
 no longer publishing)

ENVIRONMENTAL SCIENCE
 AND TECHNOLOGY
American Chemical Society
1155 16th Street, N.W.
Washington, DC 20036

FABRITE METALS CORP.
205 East 42nd Street
New York, NY 10017
(Mirrors)

FINANCIAL WORLD
Macro Publishing Corp.
17 Battery Place
New York, NY 10004

FORBES
60 Fifth Avenue
New York, NY 10011

FOUNDATION FOR THE GEN-
 ERATION OF ELECTRICITY
 BY WIND POWER
B. W. Colenbrander, Secretary
Jan Steenlaan 12
Heemstede, Netherlands

FUTURIST
World Future Society
Box 30369, Bethesda Station
Washington, DC 20014

GELIOTEKHNIKA/HELIOTECH-
 NOLOGY
Russian language journal pub-
 lished in English translation
Allerton Press
150 Fifth Avenue
New York, NY 10011

GEOLOGICAL SOCIETY OF
 AMERICA BULLETIN
3300 Penrose Avenue
Boulder, CO 80301

GEOTHERMAL ENERGY
 MAGAZINE
318 Cherrywood Street
West Covina, CA 91791

GEOTHERMAL INFORMATION
 SERVICES
318 Cherrywood Street
West Covina, CA 91791

HEALTH
American Osteopathic Association
212 East Ohio
Chicago, IL 60611

HEATING, PIPING, AND AIR
 CONDITIONING
Reinhold Publishing
600 Summer Street
Stamford, CT 06904

HELIOTECHNOLOGY/GELIO-
 TEKHNIKA see GELIO-
 TEKHNIKA

HIGH COUNTRY NEWS
140 North Seventh Street
Lander, WY 82520

HOBBIES
Lightner Publishing Co.
1006 S. Michigan Avenue
Chicago, IL 60611

HOFFMAN ELECTRONICS CORP.
El Monte, CA
(solar cells, solar-cell hand-
    books)

HOUSE AND GARDEN
Conde Nast Publications
420 Lexington Avenue
New York, NY 10017

HOUSE AND HOME
McGraw-Hill
1221 Avenue of the Americas
New York, NY 10020

HOUSE BEAUTIFUL
Hearst Magazines
717 Fifth Avenue
New York, NY 10022

HOWARD W. SAMS & CO., LTD.
4300 West 62nd Street
Indianapolis 6, IN
(Solar Cell and Photocell Exper-
    imenters Guide)

IEEE PROCEEDINGS and IEEE
    SPECTRUM
Institute of Electrical and Elec-
    tronics Engineers
345 E. 47th Street
New York, NY 10017

IMPACT OF SCIENCE ON
    SOCIETY
UNESCO
317 East 34th Street
New York, NY 10016

INDUSTRIAL DEVELOPMENT
Conway Research Inc.
2600 Apple Valley Rd.
Atlanta, GA 30319

INDUSTRIAL RESEARCH
Industrial Research, Inc.
Beverly Shores, IN 46301

INDUSTRY WEEK
Penton Publishing
Penton Bldg.
1213 W. 3rd Street
Cleveland, OH 44113

INDIAN JOURNAL OF
    ECONOMICS
University of Allahabad
Dept. of Economics and Com-
    merce
Allahabad, India

INSTITUTE OF ELECTRICAL
    AND ELECTRONICS ENGI-
    NEERS see IEEE ...

INTERNATIONAL RECTIFIER
CORP.
El Segundo, CA
(solar cells, solar-cell hand-
    books)

INTERNATIONAL SOLAR
    ENERGY SOCIETY
P.O. Box 52
Parksville, Victoria
Australia 3052
or: c/o Dr. William H. Klein
Smithsonian Radiation Biology
    Lab
12441 Parklawn Drive
Rockville, Maryland 20852

INTERNATIONAL WILDLIFE
National Wildlife Federation
1412 16th Street, N.W.
Washington, DC 20036

IRON AGE
Chilton Co.
One Decker Square
Bala Cynwyd, PA 19004

JENKINS, K. A.
Energex Corp.
5441 Paradise Road, Suite C-137
Las Vegas, NV 89119

JOURNAL OF APPLIED PHYSICS
American Institute of Physics
335 E 45th Street
New York, NY 10017

JOURNAL OF THE ELECTRO-
    CHEMICAL SOCIETY
Electrochemical Society
30 E. 42nd Street
New York, NY 10017

JOURNAL OF ENGINEERING
   FOR POWER
American Society of Mechanical
   Engineers
345 E. 47th Street
New York, NY 10017

JOURNAL OF ENVIRONMENTAL
   HEALTH
National Environmental Health
   Assoc.
1600 Pennsylvania
Denver, CO 80203

JOURNAL OF THE FRANKLIN
   INSTITUTE
Franklin Parkway
Philadelphia, PA 19103

JOURNAL OF HYDRAULICS
   DIVISION--ASCE
American Society of Civil
   Engineers
345 E. 47th Street
New York, NY 10017

JOURNAL OF POWER DIVISION
   --ASCE
(see above)

JOURNAL OF SPACECRAFT
   AND ROCKETS
American Institute of Aeronautics
   and Astronautics
1290 Avenue of the Americas
New York, NY 10019

LÖF, DR. GEORGE
Farmers Union Building
Denver, CO
(Reflector cooker)

MAAG, DR. CARL
Solarex Corporation
1335 Piccard Drive
Rockville, MD 20850
(manufactures solar equipment)

McCRACKEN, HORACE
Sunwater Company
10404 San Diego Mission Road
San Diego, CA 92108
(manufactures solar equipment)

MACHINE DESIGN
Penton Publishing Co.
Penton Plaza
1111 Chester Avenue
Cleveland, OH 44114

MARINE TECHNOLOGY SOCIETY
   JOURNAL
1730 M. Street, N.W.
Washington, DC 20036

MECHANICAL ENGINEERING
American Society of Mechanical
   Engineers
345 E. 47th Street
New York, NY 10017

MECHANIX ILLUSTRATED
Fawcett Publications Inc.
1515 Broadway
New York, NY 10036

MEMO
University of Nevada Library
Reno, NV 89507

MICRO-MO ELECTRONICS
Box 3952
Cleveland, OH 44103
(small electric motors)

MINES
Colorado School of Mines
Golden, CO 80401

MINING CONGRESS JOURNAL
American Mining Congress
1100 Ring Bldg.
Washington, DC 20036

MOSAIC (NSF)
Supt. of Documents
Washington, DC 20402

MOTHER EARTH NEWS
P.O. Box 70
Hendersonville, NC 28739

NATION
333 Avenue of the Americas
New York, NY 10014

NATIONAL GEOGRAPHIC
National Geographic Society
1145 17th Street, N.W.
Washington, DC   20036

NATIONAL OBSERVER
Dow Jones and Co.
30 Broad Street
New York, NY   10004

NATIONAL OCEANIC AND
   ATMOSPHERIC ADMINISTRA-
   TION
Office of Public Information
WSC 5   Rm 804
Rockville. MD   20852

NATIONAL PARKS AND CON-
   SERVATION
1701 18th Street, N.W.
Washington, DC   20009

NATIONAL RESEARCH
   COUNCIL OF CANADA
Building M-2
Montreal Rd.
Ottawa   K1A OR6
Ontario, Canada

NATIONAL SOCIETY OF
   PROFESSIONAL ENGINEERS
2029 K Street, N.W.
Washington, DC   20006

NATIONAL TECHNICAL INFOR-
   MATION SERVICES (NTIS)
U.S. Dept. of Commerce
5285 Port Royal Road
Springfield, VA   22151

NATURAL RESOURCES JOURNAL
University of New Mexico
   School of Law
1117 Stanford Street, NE
Albuquerque, NM   87106

NATURE
Macmillan Journals Ltd.
Little Essex St.
London WC 2R 3LF, England

NEW REPUBLIC
1244 19th Street, N.W.
Washington, DC   20036

NEW SCIENTIST
128 Long Acre
London WC2E 9QM
England

NEW YORK ACADEMY OF
   SCIENCES ANNALS
2 East 63rd Street
New York, NY   10021

NEW YORK SCIENTISTS' COM-
   MITTEE FOR PUBLIC
   INFORMATION, INC.
30 East 68th Street
New York, NY   10021

NEW YORK TIMES
229 W. 43rd Street
New York, NY   10036

NOAA   see   NATIONAL
   OCEANIC AND ATMOSPHERE
   ADMINISTRATION

NOT MAN APART
Dept. of Public Information
Pennsylvania State University
312 Old Main
University Park, PA   16802

NUCLEAR NEWS
American Nuclear Society
244 East Ogden Avenue
Hinsdale, IL   60521

THE NUCLEUS
Indiana Institute of Technology
1600 E. Washington Blvd.
Fort Wayne, IN   46803

OHIO JOURNAL OF SCIENCE
445 King Avenue
Columbus, OH   43201

OIL AND GAS JOURNAL
Petroleum Publishing Co.
211 S. Cheyenne Avenue
Box 1260
Tulsa, OK   74101

OPTICS AND SPECTROSCOPY
Optical Society of America
2100 Pennsylvania Avenue, N.W.
Washington, DC   20037

ORGANIC GARDENING AND
   FARMING
Rodale Press, Inc.
Organic Park
Emmaus, PA   18049

OUR PUBLIC LANDS
Supt. of Documents
Washington, DC   20402

OUTDOOR AMERICA
Izaak Walton League of America
1326 Waukegan Rd.
Glenview, IL   60025

PACIFIC SCIENCE
University Press of Hawaii
535 Ward Avenue
Honolulu, HI   96814

PETROLEUM ENGINEER
   INTERNA :ONAL
P. O. Box 1589
Dallas, TX   75221

PETROLEUM TECHNOLOGY
Society of Petroleum Engineers
   of AIME
6200 North Central Expressway
Dallas, TX   75206

PHYSICS TODAY
American Institute of Physics
335 East 45th Street
New York, NY   10017

PIPELINE AND GAS JOURNAL
Box 1589
Dallas, TX   75221

PLASTICS TECHNOLOGY
Bill Brothers
630 Third Avenue
New York, NY   10017

POPULAR ELECTRONICS
Ziff-Davis Publishing Co.
1 Park Avenue
New York, NY   10016

POPULAR MECHANICS
Hearst Corp.
224 W. 57th Street
New York, NY   10019

POPULAR SCIENCE
355 Lexington Ave.
New York, NY   10017

POWER
McGraw-Hill
1221 Avenue of the Americas
New York, NY   10020

POWER ENGINEERING
Technical Publishing Co.
308 East James Street
Barrington, IL   60010

POWER FROM OIL
Metropolitan Petroleum Co.
380 Madison Avenue
New York, NY   10017
(no longer publishing)

PREVENTION
Rodale Press
33 E. Minor Street
Emmaus, PA   18049

PRODUCT ENGINEERING
McGraw-Hill
1221 Avenue of the Americas
New York, NY   10020

PROFESSIONAL ENGINEER
National Society of Professional
   Engineers
2029 K. Street, N.W.
Washington, DC   20006

PROGRESSIVE ARCHITECTURE
Reinhold Publishing Co.
600 Summer St.
Stamford, CT   06904

PUBLIC POWER
American Public Power Assoc.
2600 Virginia Avenue, N.W.
Washington, DC   20037

PUBLIC UTILITIES FORT-
  NIGHTLY
Public Utilities Reports, Inc.
332 Pennsylvania Bldg.
Washington, DC 20004

QST
American Radio Relay League
225 Main Street
Newington, CT 06111

RADIO ELECTRONICS
200 Park Avenue
Suite 1101
New York, NY 10003

RAMPARTS
2054 University Avenue
Berkeley, CA 94704

READERS' DIGEST
Readers' Digest Association, Inc.
Pleasantville, NY 10570

RECLAMATION ERA
Supt. of Documents
Washington, DC 20402

RIBA JOURNAL
Royal Institute of British
  Architecture
66 Portland Pl.
London W1N 4AD, England

SATURDAY REVIEW/WORLD
488 Madison Avenue
New York, NY 10022
  see also the former monthly,
  SATURDAY REVIEW OF THE
  SCIENCES

SCIENCE
American Association for the
  Advancement of Science
1515 Massachusetts Avenue, N.W.
Washington, DC 20005

SCIENCE DIGEST
224 W. 57th Street
New York, NY 10019

SCIENCE NEWS
Science Service, Inc.
1719 N Street, N.W.
Washington, DC 20036

SCIENCE TEACHER
National Science Teachers
  Association
1201 16th Street, N.W.
Washington, DC 20036

SCIENTIFIC AMERICAN
415 Madison Avenue
New York, NY 10017

SEARCH
Exploration Division Aero
  Service Corp.
4219 Van Kirk Street
Philadelphia, PA 19135

SIERRA CLUB BULLETIN
1050 Mills Tower
San Francisco, CA 94104

SMITHSONIAN
900 Jefferson Drive
Washington, DC 20560

SOLARCELL AND PHOTOCELL
  EXPERIMENTERS GUIDE see
  HOWARD W. SAMS & CO.

SOLAR ENERGY
Pergamon Press
Maxwell House, Fairview Park
Elmsford, NY 10523

SOLAR ENERGY APPLICATIONS
  LABORATORY
Melpar, Inc.
Falls Church, VA 22042
(various solar products)

SOLAR ENERGY DIGEST
P.O. Box 17776
San Diego, CA 92117

SOLAR ENERGY RESEARCH
  AND INFORMATION CENTER
1001 Connecticut Avenue, N.W.
Washington, DC

SOLAR ENERGY SOCIETY OF
  AMERICA
P. O. Box 4264
Torrance, CA 90510

SOLAR ENERGY: The Journal
  of Solar Energy Science and
  Technology
International Solar Energy
  Society
c/o Smithsonian Radiation Biology Lab.
12441 Parklawn Dr.
Rockville, MD 20852

SOLID STATE ELECTRONICS
Pergamon Press
Maxwell House, Fairview Park
Elmsford, NY 10523

SOLAR PRODUCTS
Opa Locka, FL 33054
(solar water heaters)

SPACE WORLD
Palmer Publications
Rt. 2, Box 36
Amherst, WI

SUNDU SOLAR HEATER
c/o Mrs. Anthony J. Meagher
3319 Keys Lane
Anaheim, CA
(solar heaters)

TECHNOLOGY REVIEW
Alumni Association of M. I. T.
Cambridge, MA 02139

THOMASON, DR. HARRY E.
6802 Walker Mill Road, S. E.
Washington, DC 20027
(solar homes, solar tepee)

UMSCHAU IN WISSENSCHAFT
  UND TECHNIK
Umschau Verlag
Stuttgarter Str 18-24
6 Frankfort am Main
West Germany

UNESCO COURIER
United Nations Educational,
  Scientific and Cultural Orga-
  nization
Place de Fontenoy
75700 Paris, France

VESTNIK MASHINASTROENIYA
Gosudarstvennyi Komitet Soveta.
  Ministrov S & SS Ponauke i
  Tecknike
Prospekt Mira 106
Moscow 1-164, U. S. S. R.
(in Russian)

VITA
(Volunteers for International
  Technical Assistance)
College Campus
Schenectady, NY 12308

WALL STREET JOURNAL
Dow Jones and Company
30 Broad Street
New York, NY 10004

WATER RESOURCES BULLETIN
American Water Resources
  Association
Box 434
Urbana, IL 61802

WATER SPECTRUM
Dept. of Army, Corps of
  Engineers
Supt. of Documents
Government Printing Office
Washington, DC 20402

WORKING PAPERS IN SOCI-
  OLOGY AND ANTHROPOLOGY
University of Georgia, Athens
Dept. of Sociology and Anthro-
  pology
Athens, GA 30601

WORLD OIL
Gulf Publishing Co.
3301 Allen Parkway
P. O. Box 2608
Houston, TX 77001

WORMSER, ERIC M.
88 Foxwood Road
Stamford, CT   06903
(applied solar power)

YELLOTT ENGINEERING
  ASSOCIATES, INC.
c/o John I. Yellott
9051 North Seventh Avenue
Phoenix, AZ   85705

ZEITSCHRIFT FÜR ENERGIE-
  TECHNIK UND ENERGIE-
  WIRTSCHAFT see
  BRENNSTOFF-WÄRME-KRAFT

ZENITH SALES CORP.
1900 N. Austin Avenue
Chicago, IL   60639
(solar radios)

SOLAR ENERGY INDEX*

GENERAL	1000	1001	1003	1005	1007	1009	1010
1011	1012	1013	1014	1016	1017	1020	1025
1026	1027	1030	1031	1034	1035	1037	1041
1043	1047	1047a	1049	1050	1054	1056	1059
1060	1062	1065	1071	1072	1073	1085	1088
1091	1097	1098	1099	1100	1104	1105	1106
1107	1108	1109	1112	1114	1117a	1119	1120
1122	1124	1133	1134	1135	1137	1140	1141
1142	1143	1154	1155	1156	1157	1168	1171
1172a	1176	1181	1181a	1183	1189	1191	1194	1195
1197	1200	1208	1210	1214	1217	1220	1221
1222	1223	1224	1228	1232	1235	1243	1244
1248	1254	1256	1263	1264	1266	1268	1269
1270	1271	1272	1273	1274	1275	1276	1277
1278	1279	1280	1281	1282	1283	1284	1288
1292	1293a	1294a	1299	1313	1319	1320	1323
1324	1325	1326	1334	1335	1338	1341	1342
1343	1344	1345	1346	1350	1356	1358	1359
1361	1362	1363	1366	1367	1368	1369	1370
1371	1372	1373	1374	1375	1376	1377	1378
1381	1387	1389	1390	1392	1396	1400	1404
1405	1406	1409	1411	1413	1414	1416	1417
1419	1420	1421	1422	1423	1426a	1427a	1430
1431	1433	1438	1440	1441	1442	1444	1445
1446	1451	1460	1463	1469	1472	1480	1482
1486	1487	1488	1489	1490	1492	1500	1501
1502	1503	1504	1505		1506	1508	1512
1513a	1515	1516	1517	1519	1521	1527	1530
1534	1537	1542	1543	1544	1545	1546	1549
1553	1555	1557	1566	1567	1570	1571a	1581
1583	1586	1587	1590	1600	1605	1607	1613
1615	1616	1632	1633	1634	1635	1637	1638
1639	1640	1646	1647	1648	1651	1652	1653
1654	1664	1666	1679	1680	1681	1682	1683
1684	1685	1689	1690	1693	1694	1698	1699
1704	1710	1711	1713	1715	1717	1725	1726
1729	1730	1731	1735	1738	1739	1740	1741
1742	1744	1748	1749	1750	1751	1753	1754
1755	1756	1757	1759	1767	1769	1779	1785
1786	1789	1791	1792	1797	1798	1800	1802

*See note on page xiv

GENERAL (cont.)   1806   1807   1808   1810   1813   1818   1823
1828   1830   1832   1833   1835   1836   1838   1843   1846
1847   1848   1849   1851   1852   1857   1858   1859   1860
1868   1870   1876   1888   1889   1890   1893   1894   1898
1902   1904   1904a 1910   1912   1913   1915   1916   1918
1919   1920   1921   1922   1925   1927a 1928   1938   1939
1941   1942   1943   1945   1945a 1946   1948   1949   1950
1950a 1952   1953   1953a 1954   1955   1956   1957   1958
1960   1963   1968   1969   1971   1972   1973   1974   1975
1978   1979   1979a 1980   1980a 1982   1983   1987   1992
1993   1995   1996   1997   1998   2001   2003a 2006   2007
2008   2009   2011   2012   2013   2013a 2015   2019   2020
2021   2022   2025   2028   2029   2034   2038   2049   2050
2051   2080a 2083   2084   2095   2097   2098   2104   2105
2109   2124   2126   2128   2129   2132   2133   2134   2135
2138   2143   2145   2146   2149   2150   2153   2153a 2154
2155   2157   2161   2162   2163   2164   2165   2166   2168
2170   2171   2176   2179   2185   2192   2198   2201   2202
2203   2204   2211   2212   2220   2223   2224

BATTERIES (SOLAR CELLS)   1021   1021a 1029   1045   1046   1047
1048   1051   1064   1066   1067   1073   1075a 1095   1139
1141   1163   1172   1187   1205   1209   1218   1229   1265
1267   1287   1290   1294a 1309   1315   1322   1322a 1324
1330   1331   1348   1352   1366   1391   1393   1395   1412
1418   1428   1438   1458   1461   1462   1479   1491   1495
1496   1497   1498   1499   1513   1529   1531   1538   1550
1551   1573   1591   1592   1649   1650   1656   1692   1712
1716   1723   1724   1727   1728   1737   1761   1763   1774
1775   1819   1820   1824   1825   1839   1840   1841   1844
1855   1863   1870   1871   1874   1892   1903   1905   1906
1924   1925   1929   1930   1931   1932   1933   1934   1935
1936   1937   1947   1970   1988   1989   1991   2005   2030
2057   2082   2094   2107   2108   2110   2111   2127   2144
2147   2154   2159   2173   2175   2196   2197   2206   2213
2219   2226

COLLECTORS, CONCENTRATORS, ETC.   1001   1005   1023   1040
1069   1070   1074   1075   1076   1077   1078   1079   1080
1089   1110   1113   1116   1125   1131   1146   1149   1158
1164   1167a 1181   1190   1199   1201   1202   1213   1215
1225   1226   1230   1234   1236   1238   1239   1241   1245
1246   1247   1249   1250   1251   1252   1253   1257   1258
1259   1266   1316   1319   1327   1329   1336   1338   1339
1340   1350   1351   1354   1360   1370   1371   1388   1399
1424   1425   1427   1434   1440   1445   1448   1449   1449a
1450   1465   1466   1467   1473   1474   1475   1476   1493
1510   1511   1513   1540   1559   1560   1562   1563   1564
1576   1588   1594   1595   1596   1597   1614   1624   1631
1651   1668   1673   1694   1697   1698   1702   1703   1709

1732	1758	1773	1783	1788	1809	1824	1846	1848
1849	1850	1861	1862	1865	1871	1872	1873	1881
1882	1883	1885	1887	1895	1897	1901	1919	1944
1962	1963a	1994	1999	2006a	2023	2032	2034	2035
2036	2037	2039	2042	2044	2054	2055	2056	2078
2079	2080	2096	2099	2100	2101	2102	2121	2123
2132	2151	2153	2165	2178	2180	2181	2182	2183
2186	2188	2190	2205	2207	2209	2210	2212	2220

CONVERSION (OF HEAT TO ENERGY)					1053	1058	1092	1102
1115	1121	1127	1128	1129	1136	1166	1179	1183
1184	1185	1192	1196	1198	1227	1244	1248	1255
1288	1292	1297	1310	1311	1319	1338	1379	1380
1382	1429	1439	1447	1450	1452	1453	1464	1514
1519	1521	1525	1532	1539	1541	1547	1555	1556
1558	1587	1593	1598	1599	1625	1626	1627	1640
1642	1646	1654	1657	1658	1659	1660	1663	1678
1681	1683	1686	1691	1694	1696	1714	1722	1733
1734	1746	1755	1760	1764	1765	1770	1771	1776
1777	1778	1782	1795	1811	1814	1815	1816	1817
1821	1822	1826	1827	1829	1837	1842	1845	1854
1860	1867	1869	1877	1880	1884	1907	1942	1976
1990	1995	1996	1997	2002	2015	2048	2074	2075
2081	2110	2111	2137	2144	2153	2154	2195	2199
2200	2222							

COOKERS	1002	1060	1231	1233	1349	1407	1408	1481
1611	1618	1619	1721	1736	1790	1796	1896	1940
1990	2025a	2026	2068	2073	2076	2156	2160	

DISTILLATION		1004	1024	1052	1082	1083	1084	1177
1178	1216	1260	1314	1318	1384	1385	1386	1395
1397	1403	1453	1454	1483	1484	1489	1548	1554
1574	1582	1584	1601	1603	1687	1698	1719	1720
1794	1804	1861	1878	1886	1986	2000	2061	2062
2063	2064	2072	2128	2130	2132	2167	2194	

DRYING	1008	1068	1117	1148	1207	1608	1766	1767
1911	2047	2193	2221					

ENGINES	1050	1291	1295	1302	1303	1305	1306	1307
1308	1485	1494	1523	1526	1568	1695	2040	2041
2106	2128	2152	2225					

FURNACES	1025	1036	1055	1063	1094	1165	1167	1170
1173	1174	1180	1186	1206	1242	1304	1312	1317

AUTHOR INDEX
(Includes coauthors, editors/compilers,
and associations)

Abetti, Giorgio 1000
Abbott, C. G. 1001
Abou-Hussein, M. S. M. 1002
Abrahamson, D. E. 6000
Abrams, C. 6001, 6002
Academy Forum, National
  Academy of Sciences 1
Achenbach, P. R. 1003
Achilov, B. 1004
Adams, W. C. 1005
Ajami, F. 1061
Akyurt, M. 1008
Alberty, R. A. 1198, 1201
Alexandroff, G. 1210, 1359
Alexandroff, J. 1359
Alfven, Hannes 2, 1009
Allegretto, M. 1010
Allen, Donald R. 3000, 3001
Allison, H J. 4000
Altman, Manfred 1013, 3003
Ambrose, E. R. 1988
Ambs, L. L. 1646
American Society of Heating,
  Refrigeration and Air Condi-
  tioning Engineers 1033
Anderson, B. 1014
Anderson, D. B. 1015
Anderson, J H. 1016, 1017
Anderson, J. H., Jr. 1017
Anderson, L. B. 1018, 1019
Anderson, Richard J. 5,
  1020, 3005, 4002, 5000
Anderson, W. A. 1021, 1021a
Andiranov, V. M. 4003, 4004
Andrassy, S. 1022, 2076
Angstrom, A. 1023
Annaev, A. 1024
Aparase, A. R. 1054
Arden, Tom 3106
Armstead, H. C. H. 3007
Armstrong, Ellis L. 3008

Arnold, W. F. 1029
Aronin, J. E. 6003
Ashar, N. G. 1032, 1286
Association for Applied Solar
  Energy 1035, 1036, 1037,
  1220
Axtmann, Robert C. 1041
Ayers, D. L. 1042
Ayres, E. 1043

-B-

Backus, C. E. 1044
Bacon, F. T. 1045
Bailey, R. L. 1046
Bairamov, R. 1024
Banham, R. 6004
Barasoain, J. A. 1318
Barber, N. F. 5001
Barim, V. A. 1565
Barnes, Joseph 1047a, 3010,
  3011
Barnes, Peter 3012
Barnhart, Earle 4006
Bartz, D. R. 1048
Bassler, Friedrich 1049
Baum, V. A. 1050, 1051,
  1052, 1053, 1054, 1055
Bayless, J. R. 1056
Beck, R. 1235
Beckman, W. A. 1057, 1232,
  1402, 1553, 1873
Behrman, Dan 1059
Belanger, J. D. 1060
Bell, R. 1668
Belton, G. 1061
Bender, P. 1201
Benford, F. 1062
Benson, Arnold 4007
Benveniste, Guy 1063

200

Author Index

Werner, R.   2177
Wernick, S.   2178
Werth, Glen C.   3204
Westphal, Warren H.   3054
Whan, Glenn A.   3167
Whillier, A.   1018, 1475, 2180,
   2181, 2182, 2183
White, D. E.   3134, 3207
White, Gerald M.   2184
Whitlow, E. P.   2186
Wilcox, Carl   4138
Wilford, John Noble   3208,
   3209, 3210
Willauer, D. E.   1249
Williams, D. A.   2187, 2188,
   2189, 2190, 2191
Williams, J. Richard   2192
Williams, J. W.   1201
Willier, A.   1018, 1414, 1417
Willis, R. W.   1244
Wilson, B. W.   2193, 2194
Wilson, Howard M.   3212
Wilson, R.   6027
Wilson, V. C.   2195
Witherall, P. G.   2197
Woertz, B. B.   1476, 1477
Wohlers, H.   1661, 1662, 1663
Wohlrabe, Raymond   2198
Wolf, Martin   2199, 2200,
   2201, 2202
Wolozin, Harold   153
Wood, Samuel E.   3170
Woodall, J. M.   1479
Woodcock, G. R.   1760
Woodmill, G. W.   2203
Woodward, William   1222
Worstell, Charles C.   2205
Wujek, J. H.   2206

Zarem, A. M.   2223, 2224
Zeimer, H.   2044
Zener, C.   1581
Zmuda, J. P.   2225
Zobel, M. G. R.   3203
Zwick, E. B.   2226

-Y-Z-

Yanagimachi, M.   2208, 2209
Yannacone, Victor J.   154
Yasui, R. K.   2173
Yellot, John I.   1435, 1436,
   2210, 2211, 2212, 2213, 2214,
   2215, 2216, 2217
Yoakum, Jim   3213
Young, G. J.   2219
Young, Patrick   2220
Youngs, R. L.   2221
Yuan, E. L.   2222

KEY TO ABBREVIATIONS

ASHRAE	American Society of Heating, Refrigeration, and Air Conditioning Engineers.
IEEE	Institute of Electrical and Electronics Engineers
ASME	American Society of Mechanical Engineers
Trans.	Transactions
diagr(s)	diagram(s)
illus.	illustrated
ref(s)	reference(s)
NASA	U. S. National Aeronautics and Space Administration
NTIS	National Technical Information Service
p.	page
pp.	pages
photo(s)	photograph(s)
AIA	American Institute of Architects